AF411337

Biomechanics and Bionics in Sport and Exercise

Biomechanics and Bionics in Sport and Exercise

Editors

Rui Zhang
Wei-Hsun Tai

Basel • Beijing • Wuhan • Barcelona • Belgrade • Novi Sad • Cluj • Manchester

Editors

Rui Zhang
Key Laboratory of
Bionic Engineering
Jilin University
Changchun
China

Wei-Hsun Tai
School of Physical Education
Quanzhou Normal University
Quanzhou
China

Editorial Office
MDPI
St. Alban-Anlage 66
4052 Basel, Switzerland

This is a reprint of articles from the Special Issue published online in the open access journal *Bioengineering* (ISSN 2306-5354) (available at: www.mdpi.com/journal/bioengineering/special_issues/biomechanics_bionics).

For citation purposes, cite each article independently as indicated on the article page online and as indicated below:

Lastname, A.A.; Lastname, B.B. Article Title. *Journal Name* **Year**, *Volume Number*, Page Range.

ISBN 978-3-0365-8515-4 (Hbk)
ISBN 978-3-0365-8514-7 (PDF)
doi.org/10.3390/books978-3-0365-8514-7

Contents

Preface

In this reprint, we embark on an exhilarating exploration of the fascinating fields of biomechanics and bionics, uncovering their profound impact on human performance enhancement and the relentless pursuit of pushing the boundaries of the human body.

Biomechanics, at its core, is the study of the mechanical principles governing the movement and structure of living organisms. This scientific discipline has revolutionized our comprehension of athletic performance and injury mechanisms. Through unwavering dedication to in-depth research and groundbreaking applications, we unravel invaluable insights that enable us to optimize training methodologies, minimize injury risks, and unlock the full athletic potential that lies within the human body.

The motivation behind this scientific endeavor is driven by an unwavering passion to bridge the gap between scientific knowledge and its practical application in sports and exercise. With this in mind, the purpose of this work is to present a comprehensive compilation of groundbreaking research and technological innovations that highlight the symbiotic relationship between biomechanics and bionics in the context of sports and exercise.

This reprint is crafted with a diverse audience in mind. From seasoned researchers and scientists seeking to expand their understanding of biomechanics and bionics, to coaches, athletes, and fitness enthusiasts striving to optimize performance and prevent injuries, every reader will find valuable insights within these pages. Additionally, we hope to inspire a new generation of scholars and practitioners to join the pursuit of knowledge at the intersection of science and sports.

The collaborative efforts of esteemed authors from various disciplines, who are leaders in the field of biomechanics and bionics, have made this work possible. Their expertise and dedication have been instrumental in crafting a comprehensive resource that reflects the latest advancements and future prospects in these evolving fields.

We also extend our heartfelt acknowledgment to all the individuals and organizations whose support, assistance, and contributions have enriched this scientific endeavor. Without their collective efforts, this reprint would not have come to fruition.

Rui Zhang and Wei-Hsun Tai
Editors

 *bioengineering*

MDPI

Editorial

Cutting-Edge Research in Sports Biomechanics: From Basic Science to Applied Technology

Wei-Hsun Tai [1,2,*], Rui Zhang [2,3,*] and Liangliang Zhao [3]

[1] Graduate School, Chengdu Sport University, Chengdu 610000, China
[2] School of Physical Education, Quanzhou Normal University, Quanzhou 362000, China
[3] Key Laboratory of Bionic Engineering (Ministry of Education, China), Jilin University, Changchun 130022, China; llzhao21@mails.jlu.edu.cn
* Correspondence: dlove520@hotmail.com (W.-H.T.); zhangrui@jlu.edu.cn (R.Z.)

1. Introduction

Sports biomechanics is the study of the mechanical principles of human movement and how they apply to sports performance [1]. It involves the analysis of motion, force, and energy during sports activities and aims to understand the biomechanical factors that influence performance and injury risk [2]. Sports biomechanics is an interdisciplinary field that combines elements of engineering, physics, anatomy, and physiology to help athletes optimize their performance and reduce the risk of injury [3]. Understanding the biomechanics of sports is important because it can provide athletes with insights into how to improve their technique and training methods and develop new training methods and equipment that can help them perform at their best [4]. In addition to helping athletes improve their performance, sports biomechanics can also play a critical role in reducing the risk of injury [5]. By understanding the biomechanical factors that contribute to sports injuries, such as overuse or poor technique, coaches and trainers can develop injury prevention strategies that are tailored to the specific needs of individual athletes [1–5].

This Special Issue contains 11 studies that present new knowledge in the fields of sports biomechanics and bionic engineering. Our aim is to encourage the dissemination of this new knowledge and provide guidance to potential authors who are interested in submitting their manuscripts to our bioengineering journal. For this Special Issue, the editors, editorial board members, and editorial staff have sought highly valued research that advances scientific knowledge and will have a positive impact on sports biomechanics and bionic applications in sports. The authors cover a wide variety of important, innovative, and timely topics in the field. The themes include sports technology analysis [6–10], the mechanics of human motion [11–15], bionic applications and equipment design [16], and the mechanisms of sports injuries [12,13]. In this editorial, we will discuss the current state of sports biomechanics and the direction it is headed.

As mentioned above, sports biomechanics is crucial to athletes' success as it offers insights that allow athletes to optimize their performance, reduce the risk of injury, and develop new training methods and equipment [17–19]. Biomechanics and bionics have transformed the field by focusing on injury prevention and rehabilitation, developing personalized equipment, and utilizing computational modeling and artificial intelligence to optimize training regimens [20]. Continued investment in research is necessary to advance sports science further, develop new technologies and methodologies, and enhance athlete performance and safety. It is essential to support research in this area to ensure that the future generations of athletes can access the latest advancements in sports science and reach their full potential.

Citation: Tai, W.-H.; Zhang, R.; Zhao, L. Cutting-Edge Research in Sports Biomechanics: From Basic Science to Applied Technology. *Bioengineering* **2023**, *10*, 668. https://doi.org/10.3390/bioengineering10060668

Received: 5 May 2023
Accepted: 24 May 2023
Published: 1 June 2023

2. Application of Scientific Principles in Sports Biomechanics

Sports biomechanics is an interdisciplinary field that combines fundamental scientific principles with advanced technological tools to study the mechanics of human movement and its application in sports performance [1,2]. Basic scientific research in sports biomechanics involves the analysis of human movement, muscle and joint mechanics, neuromuscular control, the kinematics and kinetics of sports movements, and biomechanical modeling and simulation [3–5]. By understanding these biomechanical principles, researchers can identify the most efficient and effective techniques for athletes to use in their training and competition.

Applied technology is an essential component of sports biomechanics research, allowing the development and use of tools and equipment to measure and analyze human movement during sports activities [5]. Wearable sensors, motion capture systems, force plates, 3D printing, and virtual reality are just a few examples of applied technologies used in sports biomechanics. These tools provide precise measurements and data that are used to analyze and optimize human movement [6–19]. Furthermore, they enable the development of custom-fit equipment, training programs, and injury prevention strategies that are tailored to athletes' individual needs.

The significance of sports biomechanics research lies in its ability to optimize sports performance while reducing the risk of injury [1–5]. Athletes and coaches can, thus, apply biomechanics to identify the most effective training methods and equipment to use with this goal in mind [21–25]. The integration of basic science and applied technology in sports biomechanics research has led to the development of new training methods, equipment, and injury prevention strategies and has contributed to a better understanding of the biomechanical response to sports activities.

In conclusion, sports biomechanics is an interdisciplinary field that combines fundamental scientific principles with advanced technological tools to study the mechanics of human movement during sports activities. Applied technology plays a crucial role in sports science research by enabling the development and utilization of tools and equipment to measure and analyze human movement during sports activities. Via continued research and development, the field of sports biomechanics has the potential to revolutionize the way athletes train and compete, leading to optimized performance and a reduced risk of injury.

3. Application of Bionic Engineering Technology to Sports

Bionics is an interdisciplinary field that draws inspiration from nature to design and optimize artificial systems and devices [26–28]. It combines biology, engineering, and materials science to imitate the structure, function, and movement of living organisms [29,30]. This innovative and forward-looking approach produces new technologies that integrate the empirical, theoretical, and practical knowledge of biological origins [29,31,32].

Modern sports have become extremely competitive, and athletes' performance depends not only on their personal abilities and training but also on high-quality equipment and clothing to help them succeed in competitions [32]. Bionics has a wide range of research areas in sports, including biomimetic protective/assisted sportswear, biomimetic protective/assisted sports footwear, and biomimetic/assisted sports equipment. Among them, the application and development of sports footwear is the most in-depth research area [26,33]. In terms of biomimetic protective/assisted sports apparel, bionics mainly studies how to design and optimize sports apparel by imitating the materials and tissue structures of living organisms in nature to better adapt to the characteristics and needs of human movement [32].

The significance of bionics research lies in its ability to improve performance and the efficiency of sports equipment, reduce sports injuries, and enhance the sports experience. By conducting research in bionics, we can better understand the principles and adaptability of biological movement and apply them to the design of sports equipment, creating sports equipment that better fits human movement characteristics and needs [16,26,28,29,31]. The application of bionics in sports can have a profound impact beyond the field of athletics. It not only enhances athletes' performance and provides new ideas for related disciplines but also promotes the development and innovation of sports equipment, ultimately improving enterprises' competitiveness and market share. Additionally, bionics research can deepen our understanding of the mysteries of nature and life, driving the progress of science, technology, and human civilization.

In summary, the application of bionics covers various aspects of sports equipment, from design, materials, and structure to function and control, and has extensive and in-depth research. These applications not only improve performance and the user's experience but also offer new ideas and methods for the research and promotion of sports equipment.

4. Future Perspectives

The integration of sports biomechanics and bionics has the potential to transform sports and athletics by optimizing performance, reducing the risk of injury, and providing personalized training recommendations. For example, biomechanical analyses can help identify and correct flawed movement patterns that may lead to injury [12,13,25]. Bionic technologies can also provide support and protection to the musculoskeletal system, reducing the risk of injury during training and competition. Furthermore, the integration of advanced technologies into sports equipment and clothing can provide athletes with real-time data on their performance, allowing for more precise training and competition strategies [16,32,33]. However, as with any new technology, ethical considerations must be taken into account to ensure fairness and equality among athletes. Overall, the integration of sports biomechanics and bionics has the potential to significantly enhance human physical capabilities and transform the way we approach sports and physical activity.

5. Conclusions

Cutting-edge research in sports biomechanics is advancing our understanding of human movement and improving sports performance. The research provides insights into the fundamental principles of human movement and how they apply to sports performance. Applied technologies are developing new tools and techniques for measuring and analyzing sports movements, while bionic technologies are pushing the boundaries of what is possible for human performance. Together, these areas of research are shaping the future of sports biomechanics and opening up new possibilities for athletes to reach their full potential.

Author Contributions: Conceptualization, W.-H.T.; writing—original draft preparation; L.Z. and W.-H.T.; writing—review and editing, W.-H.T. and R.Z. All authors have read and agreed to the published version of the manuscript.

Conflicts of Interest: The authors declare no conflict of interest.

References

1. Hall, S.J. *Basic Biomechanics*, 3rd ed.; McGraw-Hill: Toronto, ON, Canada, 2009.
2. Hall, S.J. Kinetic Concepts for Analyzing Human Motion. In *Basic Biomechanics*; McGraw: New York, NY, USA, 2019. Available online: http://accessphysiotherapy.mhmedical.com/content.aspx?bookid=2433§ionid=19150336 (accessed on 23 April 2023).
3. Adrian, M.J.; Cooper, J.M. *Biomechanics of Sports*, 2nd ed.; McGraw-Hill: Columbia, MO, USA, 2005.
4. Knudson, D. Applying Biomechanics in Teaching and Coaching. In *Fundamentals of Biomechanics*; Springer: Cham, Switzerland, 2021. [CrossRef]
5. Teferi, G.; Endalew, D. Methods of Biomechanical Performance Analyses in Sport: Systematic Review. *Am. J. Sports Sci. Med.* **2020**, *8*, 47–52.
6. Grima, J.N.; Cerasola, D.; Sciriha, A.; Sillato, D.; Formosa, C.; Gatt, A.; Gauci, M.; Xerri de Caro, J.; Needham, R.; Chockalingam, N.; et al. On the Kinematics of the Forward-Facing Venetian-Style Rowing Technique. *Bioengineering* **2023**, *10*, 310. [CrossRef] [PubMed]
7. Ma, R.; Lam, W.-K.; Ding, R.; Yang, F.; Qu, F. Effects of Shoe Midfoot Bending Stiffness on Multi-Segment Foot Kinematics and Ground Reaction Force during Heel-Toe Running. *Bioengineering* **2022**, *9*, 520. [CrossRef] [PubMed]
8. Lu, Z.; Li, X.; Xuan, R.; Song, Y.; Bíró, I.; Liang, M.; Gu, Y. Effect of Heel Lift Insoles on Lower Extremity Muscle Activation and Joint Work during Barbell Squats. *Bioengineering* **2022**, *9*, 301. [CrossRef]

9. Gao, L.; Lu, Z.; Liang, M.; Baker, J.S.; Gu, Y. Influence of Different Load Conditions on Lower Extremity Biomechanics during the Lunge Squat in Novice Men. *Bioengineering* **2022**, *9*, 272. [CrossRef]

10. Biscarini, A. Dynamics of Two-Link Musculoskeletal Chains during Fast Movements: Endpoint Force, Axial, and Shear Joint Reaction Forces. *Bioengineering* **2023**, *10*, 240. [CrossRef]

11. Shan, G.; Zhang, X. Soccer Scoring Techniques—A Biomechanical Re-Conception of Time and Space for Innovations in Soccer Research and Coaching. *Bioengineering* **2022**, *9*, 333. [CrossRef]

12. Ren, S.; Liu, X.; Li, H.; Guo, Y.; Zhang, Y.; Liang, Z.; Zhang, S.; Huang, H.; Huang, X.; Ma, Z.; et al. Identification of Kinetic Abnormalities in Male Patients after Anterior Cruciate Ligament Deficiency Combined with Meniscal Injury: A Musculoskeletal Model Study of Lower Limbs during Jogging. *Bioengineering* **2022**, *9*, 716. [CrossRef]

13. Kang, M.; Zhang, T.; Yu, R.; Ganderton, C.; Adams, R.; Han, J. Effect of Different Landing Heights and Loads on Ankle Inversion Proprioception during Landing in Individuals with and without Chronic Ankle Instability. *Bioengineering* **2022**, *9*, 743. [CrossRef]

14. Xiang, L.; Gu, Y.; Rong, M.; Gao, Z.; Yang, T.; Wang, A.; Shim, V.; Fernandez, J. Shock Acceleration and Attenuation during Running with Minimalist and Maximalist Shoes: A Time- and Frequency-Domain Analysis of Tibial Acceleration. *Bioengineering* **2022**, *9*, 322. [CrossRef]

15. Du, Y.; Fan, Y. The Effect of Fatigue on Postural Control and Biomechanical Characteristic of Lunge in Badminton Players. *Bioengineering* **2023**, *10*, 301. [CrossRef] [PubMed]

16. Zhang, R.; Zhao, L.; Kong, Q.; Yu, G.; Yu, H.; Li, J.; Tai, W.-H. The Bionic High-Cushioning Midsole of Shoes Inspired by Functional Characteristics of Ostrich Foot. *Bioengineering* **2023**, *10*, 1. [CrossRef]

17. Sidorenko, S. Improvement of creativity via the six-step bio-inspiration strategy. *South East Eur. J. Arch. Des.* **2017**, *2017*, 1–8. [CrossRef]

18. Cheng, P.; Qu, F. Overview of the progress of sports shoes technology. In Proceedings of the 12th National Sports Biomechanic-sAcademic Exchange Conference, Taiyuan, China, 14 October 2008.

19. Lin, S.; Song, Y.; Cen, X.; Bálint, K.; Fekete, G.; Sun, D. The Implications of Sports Biomechanics Studies on the Research and Development of Running Shoes: A Systematic Review. *Bioengineering* **2022**, *9*, 497. [CrossRef] [PubMed]

20. Claudino, J.G.; de Oliveira Capanema, D.; De Souza, T.V.; Serrão, J.C.; Pereira, A.C.M.; Nassis, G.P. Current Approaches to the Use of Artificial Intelligence for Injury Risk Assessment and Performance Prediction in Team Sports: A Systematic Review. *Sports Med.-Open* **2019**, *5*, 28. [CrossRef] [PubMed]

21. Cushion, C. Modelling the complexity of the coaching process. *Int. J. Sports Sci. Coach.* **2007**, *2*, 395–401. [CrossRef]

22. Memmert, D.; Raabe, D. *Data Analytics in Football: Positional Data Collection, Modelling and Analysis*; Routledge: Abingdon, UK, 2018.

23. McGarry, T.; O'Donoghue, P.; Sampaio, J. *Handbook of Sports Performance Analysis*; Routledge: Abingdon, UK, 2013.

24. Tai, W.-H.; Yu, H.-B.; Tang, R.-H.; Huang, C.-F.; Wei, Y.; Peng, H.-T. Handheld-Load-Specific Jump Training over 8 Weeks Improves Standing Broad Jump Performance in Adolescent Athletes. *Healthcare* **2022**, *10*, 2301. [CrossRef]

25. Lin, J.-Z.; Lin, Y.-A.; Tai, W.-H.; Chen, C.-Y. Influence of Landing in Neuromuscular Control and Ground Reaction Force withAnkle Instability: A Narrative Review. *Bioengineering* **2022**, *9*, 68. [CrossRef]

26. Yu, H.-B.; Zhang, R.; Yu, G.-L.; Wang, H.-T.; Wang, D.-C.; Tai, W.-H.; Huang, J.-L. A New Inspiration in Bionic Shock Absorption Midsole Design and Engineering. *Appl. Sci.* **2021**, *11*, 9679. [CrossRef]

27. Lurie, E. Product and technology innovation: What can biomimicry inspire? *J. Biotechnol. Adv.* **2014**, *32*, 1494–1505. [CrossRef]

28. Helms, M.; Vattam, S.S.; Goel, A.K. Biologically inspired design: Process and products. *Des. Stud.* **2009**, *30*, 606–622. [CrossRef]

29. Lotfabadi, P.; Alibaba, H.Z.; Arfaei, A. Sustainability; as a combination of parametric patterns and bionic strategies. *Renew. Sustain. Energy Rev.* **2016**, *57*, 1337–1346. [CrossRef]

30. Chen, H.; Zhang, P.; Zhang, L.; Liu, H.; Jiang, Y.; Zhang, D.; Han, Z.; Jiang, L. Continuous directional water transport on the peristome surface of Nepenthes alata. *Nature* **2016**, *532*, 85–89. [CrossRef]

31. Aziz, M.S. Biomimicry as an approach for bio-inspired structure with the aid of computation. *Alex. Eng. J.* **2016**, *55*, 707–714. [CrossRef]

32. Nazemi, S.; Khajavi, R.; Bagherzadeh, R. CFD and experimental studies on drag force of swimsuit fabric coated by silica nano particles. *J. Text. Inst.* **2023**, *114*, 43–54. [CrossRef]

33. Ramirez-Bautista, J.A.; Huerta-Ruelas, J.A.; Chaparro-Cardenas, S.L.; Hernandez-Zavala, A. A Review in Detection and Monitoring Gait Disorders Using In-Shoe Plantar Measurement Systems. *IEEE Rev. Biomed. Eng.* **2017**, *10*, 299–309. [CrossRef] [PubMed]

 *bioengineering*

Article

On the Kinematics of the Forward-Facing Venetian-Style Rowing Technique

Joseph N. Grima [1,2,*], Dario Cerasola [3,4], Anabel Sciriha [5], Darren Sillato [6], Cynthia Formosa [6,7], Alfred Gatt [6,7], Michael Gauci [1], John Xerri de Caro [5], Robert Needham [7], Nachiappan Chockalingam [6,7] and Tonio P. Agius [5,*]

[1] Metamaterials Unit, Faculty of Science, University of Malta, MSD 2080 Msida, Malta
[2] Siġġiewi Rowing Club, 181, Melita Street, VLT 1129 Valletta, Malta
[3] Italian Rowing Federation, Viale Tiziano, 74, 00196 Rome, Italy
[4] Department of Psychology, Educational Science and Human Movement, University of Palermo, 90100 Palermo, Italy
[5] Department of Physiotherapy, Faculty of Health Sciences, University of Malta, MSD 2080 Msida, Malta
[6] Department of Podiatry, Faculty of Health Sciences, University of Malta, MSD 2080 Msida, Malta
[7] Centre for Biomechanics and Rehabilitation Technologies, School of Health, Science and Wellbeing, Staffordshire University, Stoke-on-Trent ST4 2DF, UK
* Correspondence: joseph.grima@um.edu.mt (J.N.G.); tonio.p.agius@um.edu.mt (T.P.A.)

Abstract: This work presents a qualitative and quantitative pilot study which explores the kinematics of Venetian style forward-facing standing rowing as practised by able-bodied competitive athletes. The technique, made famous by the gondoliers, was replicated in a biomechanics laboratory by a cohort of four experienced rowers who compete in this style at National Level events in Malta. Athletes were marked with reflective markers following the modified Helen Hayes model and asked to row in a manner which mimics their on-water practise and recorded using a Vicon optoelectronic motion capture system. Data collected were compared to its equivalent using a standard sliding-seat ergometer as well as data collated from observations of athletes rowing on water, thus permitting the documentation of the manner of how this technique is performed. It was shown that this rowing style is characterised by rather asymmetric and complex kinematics, particularly upper-body movements which provides the athlete with a total-body workout involving all major muscle groups working either isometrically, to provide stability, or actively.

Keywords: rowing; venetian rowing; kinematics; sports biomechanics; traditional rowing; motion analysis

Citation: Grima, J.N.; Cerasola, D.; Sciriha, A.; Sillato, D.; Formosa, C.; Gatt, A.; Gauci, M.; Xerri de Caro, J.; Needham, R.; Chockalingam, N.; et al. On the Kinematics of the Forward-Facing Venetian-Style Rowing Technique. *Bioengineering* **2023**, *10*, 310. https://doi.org/10.3390/bioengineering10030310

Academic Editors: Andrea Cataldo and Aurélien Courvoisier

Received: 6 December 2022
Revised: 11 January 2023
Accepted: 24 February 2023
Published: 28 February 2023

1. Introduction

Rowing, one of the oldest sports that form a part of the Olympic program, is generally practised with the athlete seated facing the rear of the direction of travel. Other rowing styles exist, such as the mode popularised by the gondoliers in Venice in which the athlete is not seated, but standing and forward facing [1,2]. The latter style is widely practised as a competitive sport in various regions of Italy, where it is referred to as "*Voga Veneta*", "*Voga alla Veneziana*" (Venetian rowing) or "*Voga in Piedi*" (standing rowing), and in other parts of the world. Compared to classic "traditional" rowing where the rower sits on fixed-seat boats, or the more biomechanically efficient "Olympic-type rowing" where the seat is allowed to slide, standing rowing offers the advantage that the athletes can easily steer the boat without needing to look backwards, or to make use of a coxswain. This makes it considerably safer to practise, particularly in harbours or similarly busy waters, and permits the athletes to focus on rowing as an exercise rather than concentrating on how to avoid collisions.

A European nation where standing Venetian-style rowing is widely practised as a sport is the Republic of Malta, a country neighbouring Italy. Here, a historical National Regatta rowed on traditional wooden boats over a distance of slightly over 1 km, which traces its origin to the 17th century, sees most of the competing boats rowed by a combination of sitting and standing rowers. Such crews are composed of two or four athletes where, as shown in Figure 1a, the standing rowers stand on the rear part of the boat, facing the sitting rowers who sit on fixed seats in the front. The basic movements of this technique, illustrated in Figure 1b, require standing rowers to essentially "push" the oar in the active "drive phase", from the "catch" to the "finish", in synchrony with the "pulling" actions of seated rowers (and vice versa in the "recovery phase").

Figure 1. The basics of Venetian style standing rowing, demonstrated here on the Maltese *"Dgħajsa tal-Pass"* racing traditional rowing boat where the standing rower is shown alongside a seated rowing. (a) A shows the "finish" position for the boat set up with a crew of two (**left**) or four (**right**) rowers whilst (b) highlights the key components of the cycle, where a standing rower (locally known as the *"parasija"*) is rowing in sync with a standard traditional seated rower (locally known as *"irmiġġier"*). Illustrated here are, approximately, (**i**) the finish, (**ii**) mid-recovery, (**iii**) the catch and (**iv**) mid-drive. Note that in boats with two rowers, the seated rower keeps the stoke rate whilst the standing rower steers.

Although traditional seated rowing, Olympic-style rowing as well as indoor rowing (i.e., rowing on an ergometer (erg), a simulated form of rowing used for training and competitions, which doubles as a very useful tool for studying sliding-seat rowing within a controlled laboratory setting) have been studied extensively from various perspectives [3,4], ranging from biomechanics [5–14], physiology [15,16], strategy [17–21], and injuries [22–28], the scientific literature regarding the sporting practise of Venetian-style standing rowing remains scarce [1,2], particularly from the aspect of biomechanics, despite the long tradition and the relative popularity of this sport. This work aims to address this lacuna by documenting, for the first time, the rather complex kinematics of the forward-facing Venetian Rowing Technique (VRT), which have been recorded using calibrated motion-capture equipment in a laboratory setting.

2. Materials and Methods

2.1. Participants

Four oarsmen (referred to as Rower 1, 2, 3 and 4), all male, having an average age of 24.5 ± 2.6 years, average height 180.0 ± 8.0 cm, average weight 94.5 ± 11.0 kg and average body mass index (BMI) of 30.0 ± 3.7, who row in the Maltese National Regatta in the standing *parasija* positions (i.e., in Venetian-style, with the oar on their right), volunteered to participate in this study. Rowers 1 and 3 were of similar height and BMI, taller than the other two rowers who also had similar height but very different BMI (Rower 2 had BMI of 26, Rower 4 had BMI of 35, compared to Rowers 1 and 3 who had a BMI of 29 and 30 respectively). The chosen inclusion criteria for this study were that the participants:

i. had to have been rowing for at least three years in this form;
ii. were above the age of 18;
iii. did not suffer from any acute musculoskeletal injuries or other comorbidities e.g., cardiorespiratory issues.

All participants fulfilled these criteria and exceeded them in terms of rowing experience. The four participating rowers had been rowing since their childhood, were regular participants in the Maltese National Regattas at the time of the study and had won several medals at national level in their rowing careers.

Participants were pre-informed about the procedure verbally and through a detailed information letter, and voluntarily agreed to participate in the study. All participants were informed that the procedure would be non-invasive and would pose no different risk to those they were accustomed to in their sport training. Their intention to participate in this study was signified through signing of an informed consent form that was previously approved by the Faculty Research and Ethics Committee (Approval number: FREC FHS_1718_017). Informed consent was also obtained to publish the information/images in an online open access publication.

2.2. Protocol Used

Standing Venetian-style rowing, as practised by athletes in Malta, was simulated using a standard Concept2 Model D rowing ergometer modified as in Figure 2a to be able to replicate Venetian-style standing with the oar on the right-hand side of the rower. This setup was designed and constructed in a manner which closely mimicked the boat's interior, both in terms of size and rowing experience. Key to this design were: (i) the ergometer itself was able to reproduce a sensation of resistance as felt by the athletes when rowing on-water; (ii) a vertical pole with a horizontal stopper, which was set at the appropriate height and separation from the rower to replicate the oar-lock position; (iii) a wooden stick to replicate the oar, replacing the ergometer handle as shown in Figure 2a, which was held by the rower on one side, and affixed near its other end to the aforementioned oar-lock replica; (iv) a rope which was knotted in a manner which replicated the normal oar-lock knot used on boats. Note that in this case, since Venetian rowing is essentially a "pushing" action, rather than "pulling", as is the case in conventional seated rowing, the ergometer had to face in the opposite direction the rower faced. Moreover, the ergometer chain had to be attached to the oar and positioned between the rower and the oar-lock, such that it did not interfere with the rower's movements.

Data collection was conducted in the Biomechanics Research Laboratory at the Faculty of Health Sciences, University of Malta using a Vicon optoelectronic motion-capture system (Oxford metrics, Oxford, UK) comprised of 16-cameras hung on solid brackets below the ceiling, operating at a sample rate of 100 Hz. This "Optical-Passive" setup could record the position of retroreflective markers placed on individuals that are tracked using 16 infrared cameras, which were calibrated using a standard calibration protocol. Athletes' movements were recorded inside this calibrated volume and the Vicon Nexus® software version 2.8.1 (Oxford metrics, Oxford, UK) was used to record the coordinates and to generate a set of angular measurements related to the joints performing the movements.

On the day of the experiments, participants were first asked to familiarise themselves with the laboratory setup so that they could give feedback on its ability to adequately replicate the real boat setup. They were then marked with 39 spherical reflective markers following the modified Helen Hayes model (PlugIn Gait, Vicon, Oxford, UK), see Figure 2(a-ii), the position of which was detected using the 16-camera setup, and asked to test the setup again and warm up in their preferred manner.

Participants were then asked to row on this setup at a self-selected power and stroke rate (fairly common in gait analysis [29]), and the three-dimensional movements of the trunk, pelvis, upper and lower limbs were sampled over four 10 s capture with these being recorded at randomly selected periods during the experiment. Data extracted from this protocol included:

(a) Angles which measure the orientation of the thorax (Thorax1; Thorax2 and Thorax3) and pelvis (Pelvis1; Pelvis2 and Pelvis3) relative to the global axis, as well as spine angles (Spine1; Spine2 and Spine3) which relate with the measurements of the thorax and pelvis angles, relative to each other;
(b) Angles which relate to the lower limbs, namely the hip (Hip1; Hip2 and Hip3), knee (Knee1) and ankle (Ankle1) angles;
(c) Angles which relate to upper limbs, namely the shoulder angles (Shoulder1; Shoulder 2 and Shoulder 3) and Ankle (Ankle 1) angles.

where the numerals '1', '2' and '3' refer to the measurement being in the sagittal, coronal, and transverse plane, or its equivalent.

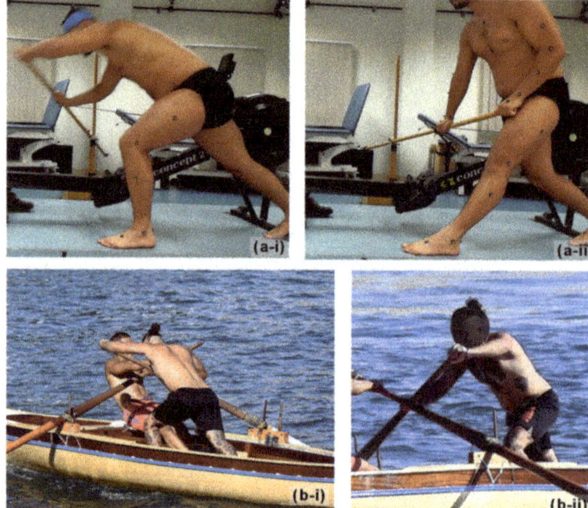

Figure 2. (**a**) The set up used in the laboratory to mimic the on-water setup, shown in (**b**). Note that the rowers in (**a-i**) and (**b**) are nearing to the "finish" whilst the rower in (**a-ii**) is nearing the "catch". Note also the position of the reflective markers, shown more clearly in (**a-ii**).

2.3. Analysis

Three full rowing cycles (strokes), starting from the "finish" (i.e., from the end of the drive phase / start of the recovery), for each of the four rowers (i.e., twelve cycles in total) were selected for further analysis and synchronised in a manner that the "catch" and finish" were made to correspond and then plotted as a percentage of the cycle.

Moreover, for each of these twelve standing rowing cycles, and for each angular measurement (θ made, the following key data were noted and further analysed statistically to obtain the mean and standard deviation for each angular parameter:

θ_{max} the maximum value of the angle θ attained within the cycle
θ_{min} the minimum value of the angle θ attained within the cycle
θ_{ROM} the range of motion associated with the angular measurement θ where $\theta_{ROM} = \theta_{max} - \theta_{min}$

This data were also compared to equivalent data for rowing on a standard sliding-seat ergometer obtained from a cohort of nine rowers, total number of rowing cycles analysed $n = 3 \times 8 = 24$ (average age: 23.9 ± 2.6 years, average height 177.6 ± 6.8 cm, average weight 89.1 ± 9.9 kg, all fixed-seat competitive rowers who also trained and/or competed using the standard sliding-seat rowing ergometer.). Analyses through Inferential statistics using the Statistical Package for Social Science (SPSS®) software, version 25.0 (IBM Corp., Released 2017, Armonk, NY: IBM Corp.), was carried out on these angular parameters using the Mann–Whitney U-test. This test was chosen as a test for normality of the parameters $\theta_{\max}$, $\theta_{\min}$ and θ_{ROM} using the Shapiro–Wilk test suggested that the data were not normally distributed.

2.4. Comparison with On-Water Rowing

Apart from the data collected in the laboratory, additional data were also collected from on-water rowing during training in the period just before a regatta. Rowers were observed in a non-intrusive manner, to negate the well-known Hawthorne and Rosenthal effects, from a public vantage point, located in the second part of the normal racing route, that is typically reserved for spectators and photographers. The rowing action was recorded using a digital camera (Nikon D3200 DSLR camera with a Sigma lens (150–600 mm), or a Nikon DSLR Camera D750, with a Sigma 70–200 f/2.8) at a high enough zoom-level where the movements of individuals were clearly and sharply identifiable, ensuring availability of around two to four cycles to have at least one complete cycle for analysis. Recording was performed 2–3 m above sea-level with the camera pointed perpendicular to the racing route to minimise perspective or parallax errors. Images were extracted from the respective clip at an appropriate frame rate and manually reviewed to choose the optimal captures. The coordinates of key points which described the actions of the rowers were identified from these captures through an in-house written script which outputs the pixel numbers of selected positions. Measurements from these coordinates were then used to verify the shape of the trends obtained quantitatively in the laboratory, e.g., findings related to trunk orientation.

2.5. Reporting

The kinematic data collected in the laboratory were reported through a combination of sequences of images, as well as plots which provides a graphical report of the manner of how the angular parameters vary, from finish to finish as a percentage of the cycle, analogous to earlier work related to standard (sliding-seat ergometer) rowing [12] and gait analysis (reported as a percentage of gait cycle) [30]. Note that in these plots these angular parameters were reported as the averaged results ± 1.96 standard deviations to provide an indication of the spread of data points, even if, given the small sample size of only four participants in the standing rower study, the assumption of normality of the data may not be true. In fact, a test for normality of the parameters $\theta_{\max}$, $\theta_{\min}$ and θ_{ROM} using the Shapiro–Wilk test suggested that the data were not normally distributed. Various angle–angle graphs are also plotted to help provide a better appreciation of the kinematics of standing rowing.

3. Results

A sequence of photographic images showing a typical full stroke (rowing cycle) of a standing (*parasija*) and sitting athlete (*irmiġġier*) rowing the traditional *Tal-Pass* Maltese boat are shown in Figure 3, whilst a second sequence of photographic images showing a typical subject undergoing this experiment in the laboratory are shown in Figure 4.

Examining first the on-water sequence of images, it should first and foremost be noted that standing Venetian style rowing cycle can also be described in terms of an active "drive" phase (Figure 3f–l) and a "recovery" phase (Figure 3a–e), which is essentially a reverse of the drive. Given that, as on any boat, the "catch" and "finish" of the two rowers need to be perfectly synchronised timewise to ensure the boat moves efficiently, one can look at the movement of the sitting rower to help interpret the kinematics of the standing rower, a feature which makes Malta an ideal location for this present study. As evident in these images, the standing rower "pushes" the oar in the active drive phase, accompanying the "pulling" action of the sitting rower, to take the oar from the "catch" (Figure 3f) to the "finish" (Figure 3l/Figure 3a), whilst in the recovery phase, the standing rower "pulls" the oar, accompanying the "pushing" action of the sitting rower. Of note is that a coxswain was not required as the standing rower faces the forward direction takes the role to steer the boat.

A comparison of images in Figure 3 with those in Figure 4, both involving a different subject, suggests that the on-water kinematics were very well replicated in the laboratory setup. Figure 4 depicts the "recovery" phase in the first two rows of images and the "drive" phase in the last two rows.

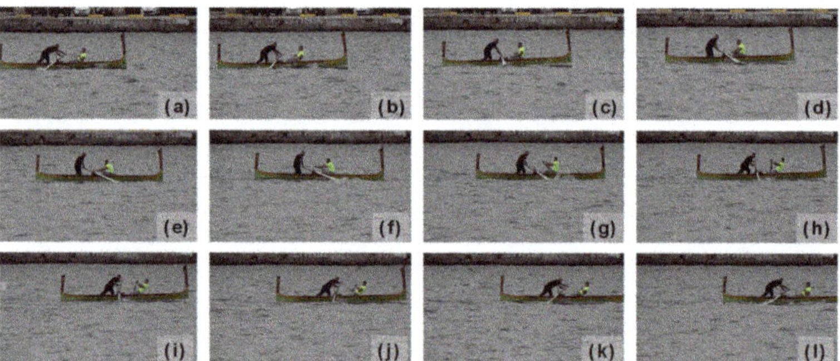

Figure 3. A detailed sequence of images showing the kinematics of standing rowing recorded on water on a Maltese *Dgħajsa tal-Pass*. Note that the catch, drive, finish and recovery of the sitting and standing rowers is synchronised, as is normal in rowing. (**a–l**) illustrate different parts of the stroke with (**a,l**) representing the finish; with the catch commencing at (**e**).

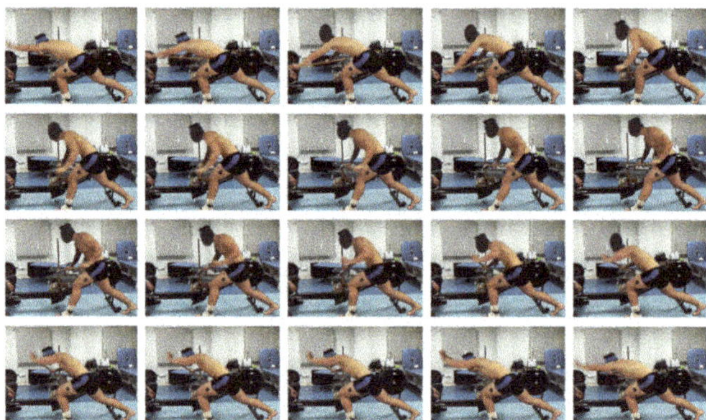

Figure 4. A sequence of images showing a typical standing rowing cycle as performed in the lab, with the first two rows of images showing the "recovery" phase (starting from the "finish" position) and the last two rows showing the "drive" phase (starting from the "catch", ending with the "finish").

Moreover, as illustrated in these two sets of images (Figures 3 and 4), a number of qualitative inferences, which are discussed in more detail in the next section, could be made, including:

i.   The kinematics of standing rowing are rather asymmetric, and involve complex movements, particularly upper-body movements;

ii.   Standing rowing provides the athlete with a total-body workout involving all major muscle groups working either isometrically, to provide stability, or actively;

iii.   The energy input of the sitting and standing rower to propel the boat seem, to a first approximation, to be not too dissimilar from each other, as discussed further below.

Moreover, quantitative results of angular measurements taken in the laboratory are reported as graphs showing the variation of the angular measurements with a percentage of the cycle, finish to finish and in tables which report the mean and standard deviations of θ_{ROM}, θ_{min} and θ_{max}. These measurements, as discussed in more detail below, confirm the complexity and asymmetricity of standing rowing, which is very distinct from that of standard seated rowing. The tables presented contain information related to each of the four rowers studied, the average of the four rowers, and a comparison with the respective average values measured for standard sliding-seat ergometer rowing (together with the respective p-values comparing the average measurements for standing rowing and standard sliding-seat ergometer rowing).

3.1. Measurements Related to the Kinematics of the Thorax, Pelvis and Spine

Measurements of the global orientation of the thorax, pelvis, and of the spine relative to the pelvis for standing rowing are reported in Figure 5, which provides a graphical report of how the measurements of the angular parameters vary, from finish to finish. These angular parameters are reported as averaged results ± 1.96 standard deviations (where the value of ± 1.96 should not be taken as implying that the data are normally distributed). This figure also compares this data with its equivalent related to standard ergometer rowing (the control group). To facilitate interpretation, Figure 6 shows a re-plot of the sagittal plots in Figure 5, in the format of angle–angle plots, where the pelvis and spine angular measurements are plotted against the thorax angles. A more complete set of the angle–angle graphs is shown in the Supplementary Materials.

In addition, reported in Tables 1–3 are the ranges of motion θ_{ROM}, the minimum and maximum values ($\theta_{\min}$ and $\theta_{\max}$) of each angular measurement plotted in Figure 5, together with the p-values related to a comparison of the average angular parameters. Note that only angles related to the left side of the body are being reported as the measurements of the angles in the sagittal, transverse, and coronal planes are related through symmetry between the left and right (equality for sagittal movements; mirror image for the transverse and coronal planes), as expected.

Table 1. The range of motion values, θ_{ROM}, of the thorax, pelvis and spine angles measured for the Venetian style rowing technique, in degrees, $\pm$ 1 s.d., for the individual rowers 1–4, their mean, and its equivalent mean value for standard ergometer rowing, with p-value for these two means reported in parentheses.

θ_{ROM}		Venetian Rowing Technique (VRT)					Control–Erg	
		Rower 1	Rower 2	Rower 3	Rower 4	Mean	Mean	(*p*)
Thorax1	L	24.7 ± 3.1	47.7 ± 1.2	39 ± 4.4	18 ± 3	32.3 ± 12.5	52 ± 11.3	(0.000)
Thorax2	L	12 ± 1	17 ± 1	21.3 ± 2.5	9.3 ± 2.9	14.9 ± 5.1	2.9 ± 1.2	(0.000)
Thorax3	L	10.7 ± 1.5	11.7 ± 0.6	20.3 ± 4.2	15.3 ± 2.9	14.5 ± 4.6	3.3 ± 1.6	(0.000)
Pelvis1	L	29.3 ± 2.9	44 ± 2	30.3 ± 1.5	18 ± 5.6	30.4 ± 10	36.3 ± 9.3	(0.139)
Pelvis2	L	12 ± 1.7	10 ± 2	21.7 ± 1.5	9.7 ± 2.1	13.3 ± 5.3	2.7 ± 1.5	(0.000)
Pelvis3	L	10.3 ± 2.1	13 ± 1	20.7 ± 3.8	15 ± 3.6	14.8 ± 4.7	3.5 ± 1.6	(0.000)
Spine1	L	9.3 ± 1.5	7.3 ± 1.5	7.3 ± 2.1	4 ± 1	7 ± 2.4	17.3 ± 6.6	(0.000)
Spine2	L	6.3 ± 0.6	5.7 ± 1.5	6.7 ± 1.2	5.7 ± 2.1	6.1 ± 1.3	2.8 ± 0.9	(0.000)
Spine3	L	7.3 ± 1.5	5.3 ± 1.5	4.7 ± 1.2	4.7 ± 1.2	5.5 ± 1.6	2.9 ± 1	(0.000)

Table 2. The minimum values, $\theta_{\min}$, of the thorax, pelvis and spine angles measured for the Venetian style rowing technique, in degrees, $\pm$ 1 s.d., for the individual rowers 1–4, their mean, and its equivalent mean value for standard ergometer rowing, with p-value for these two means reported in parentheses.

$\theta_{\min}$		Venetian Rowing Technique (VRT)					Control–Erg	
		Rower 1	Rower 2	Rower 3	Rower 4	Mean	Mean	(*p*)
Thorax1	L	46.7 ± 3.2	35 ± 1.7	37 ± 2.6	39.3 ± 2.1	39.5 ± 5.1	-20 ± 7.6	(0.000)
Thorax2	L	3.3 ± 1.5	0 ± 1	-7.3 ± 2.5	-7.3 ± 1.5	-2.8 ± 5.1	-1.4 ± 2.2	(0.427)
Thorax3	L	-11.3 ± 0.6	-12 ± 1.7	-16.3 ± 3.5	-14 ± 1.7	-13.4 ± 2.7	-1.5 ± 1.3	(0.000)
Pelvis1	L	15.7 ± 2.5	11.7 ± 3.2	24.3 ± 1.5	26 ± 5.3	19.4 ± 6.9	-48.3 ± 4.7	(0.000)
Pelvis2	L	7.3 ± 1.2	-4.3 ± 1.2	-3.7 ± 0.6	-3.3 ± 2.5	-1 ± 5.2	-2.1 ± 1.6	(0.000)
Pelvis3	L	-8.3 ± 1.2	-8.7 ± 0.6	-17.7 ± 3.5	-7 ± 1	-10.4 ± 4.7	-2.8 ± 1.7	(0.000)
Spine1	L	24.3 ± 1.2	21.3 ± 2.5	13.3 ± 1.2	11.3 ± 2.5	17.6 ± 5.9	27.4 ± 8.6	(0.002)
Spine2	L	-0.3 ± 2.1	-11.3 ± 2.9	-0.7 ± 1.5	0.7 ± 2.5	-2.9 ± 5.5	-2.3 ± 1.9	(0.397)
Spine3	L	-6.3 ± 2.3	-2.7 ± 0.6	-4.7 ± 0.6	4 ± 1	-2.4 ± 4.3	-2.6 ± 2.6	(0.577)

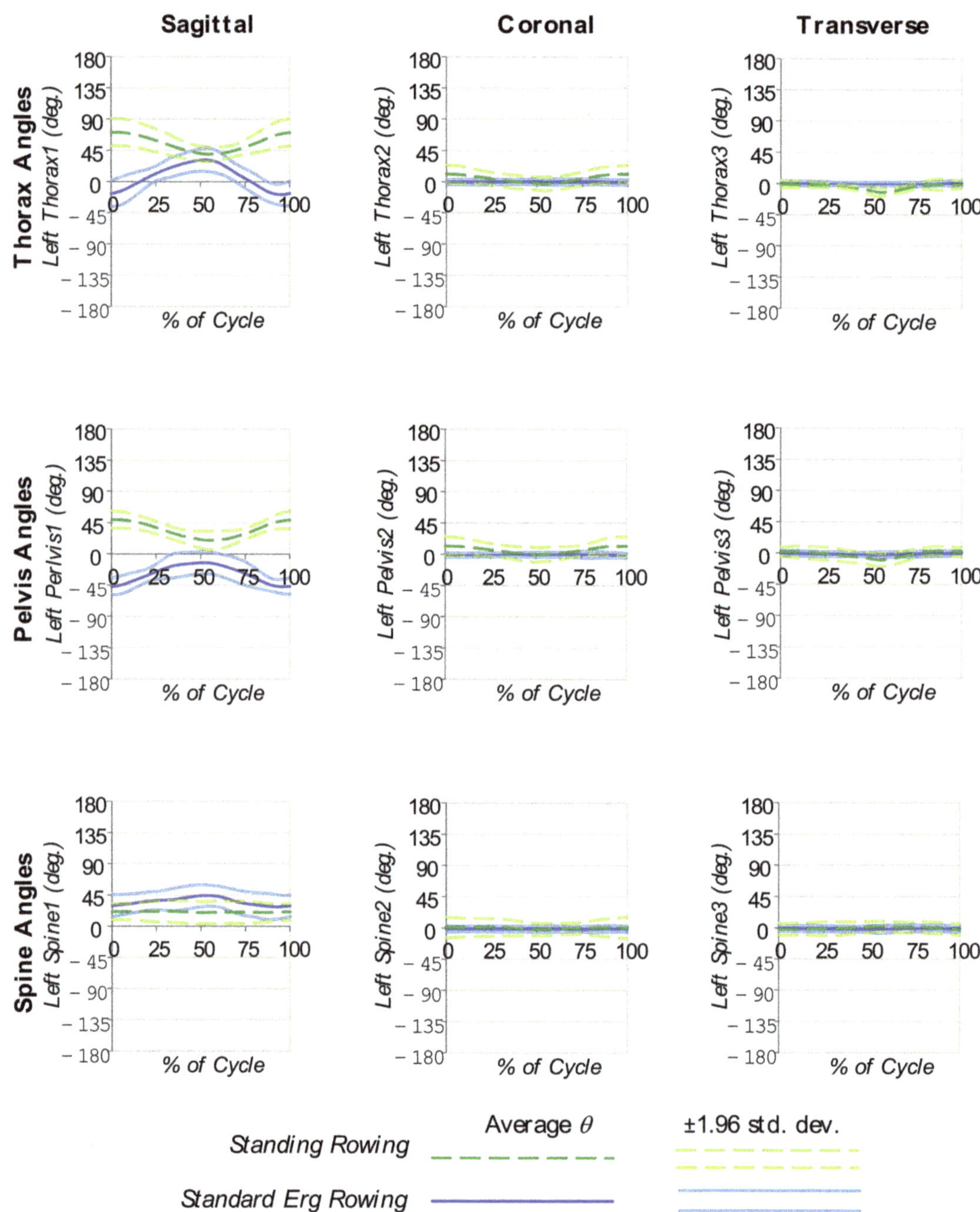

Figure 5. The averaged angular measurements for the thorax, pelvis and spine from the laboratory study, reported as mean ± 1.96 s.d., which compares standing rowing (green) with standard ergometer sliding-seat rowing (blue).

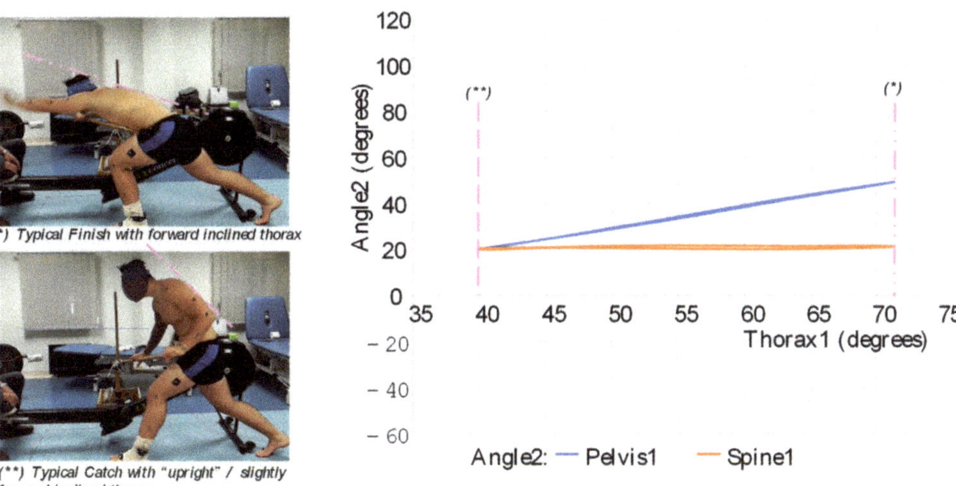

Figure 6. Angle–Angle plots, which illustrate the relationship between movements in sagittal plane of the pelvis/spine with that of the thorax. See also Supplementary Materials.

Table 3. The maximum values, $\theta_{\max}$, of the thorax, pelvis and spine angles measured for the Venetian style rowing technique, in degrees, $\pm$ 1 s.d., compared to its equivalent mean value for standard ergometer rowing.

$\theta_{\max}$		Venetian Rowing Technique (VRT)					Control–Erg	
		Rower 1	Rower 2	Rower 3	Rower 4	Mean	Mean	(*p*)
Thorax1	L	71 ± 1	82.7 ± 1.5	76 ± 2	57 ± 3.5	71.7 ± 10	32.2 ± 8.5	(0.000)
Thorax2	L	15.3 ± 2.5	16.7 ± 0.6	14.3 ± 0.6	2 ± 2	12.1 ± 6.3	1.7 ± 1.8	(0.000)
Thorax3	L	-0.3 ± 0.6	-0.3 ± 1.2	4 ± 1	1.7 ± 2.5	1.3 ± 2.3	1.8 ± 1.7	(0.314)
Pelvis1	L	45 ± 1.7	55.3 ± 4.7	55 ± 1	43.7 ± 0.6	49.8 ± 6.1	-12 ± 8.3	(0.000)
Pelvis2	L	19.7 ± 2.1	5.3 ± 2.1	18 ± 2	6.7 ± 0.6	12.4 ± 6.9	0.8 ± 2	(0.000)
Pelvis3	L	2 ± 3	4.3 ± 1.5	3 ± 1.7	8.3 ± 4	4.4 ± 3.4	0.7 ± 2	(0.001)
Spine1	L	33.7 ± 2.1	29 ± 2.6	20.3 ± 2.5	15 ± 3.6	24.5 ± 8	44.8 ± 7.8	(0.000)
Spine2	L	6 ± 1.7	-6.3 ± 1.2	5.7 ± 2.1	6.3 ± 1.5	2.9 ± 5.8	0.4 ± 2.2	(0.015)
Spine3	L	0.3 ± 1.2	2.7 ± 1.5	0 ± 1	9 ± 1	3 ± 3.9	0.3 ± 2.6	(0.061)

Moreover, Figure 7 reports the traced outlines of the back profiles for various standing rowers whilst rowing on-water, which are meant to be treated as semi-quantitative estimates to support the quantitative results reported above.

3.2. Measurements Related to Kinematics of the Knee, Ankle and Hip Joints

Measurements related to the hip, knee and ankle are reported in Figures 8 and 9. These plots report the variation, from finish to finish, of the angular parameters for the standing rowers as averaged results ±1.96 standard deviations, together with a comparison for standard ergometer rowing. For the sagittal plane, these data are also re-plotted, in the form of angle–angle plots, in Figure 10 to highlight the relationships between the various angular measurements and the inclination of the thorax. Moreover, Tables 4–6 also report θ_{ROM}, $\theta_{\min}$ and $\theta_{\max}$ for each angular measurement plotted in Figures 8 and 9, together with the corresponding *p*-values related to a comparison of standing rowing with standard ergometer sliding-seat rowing.

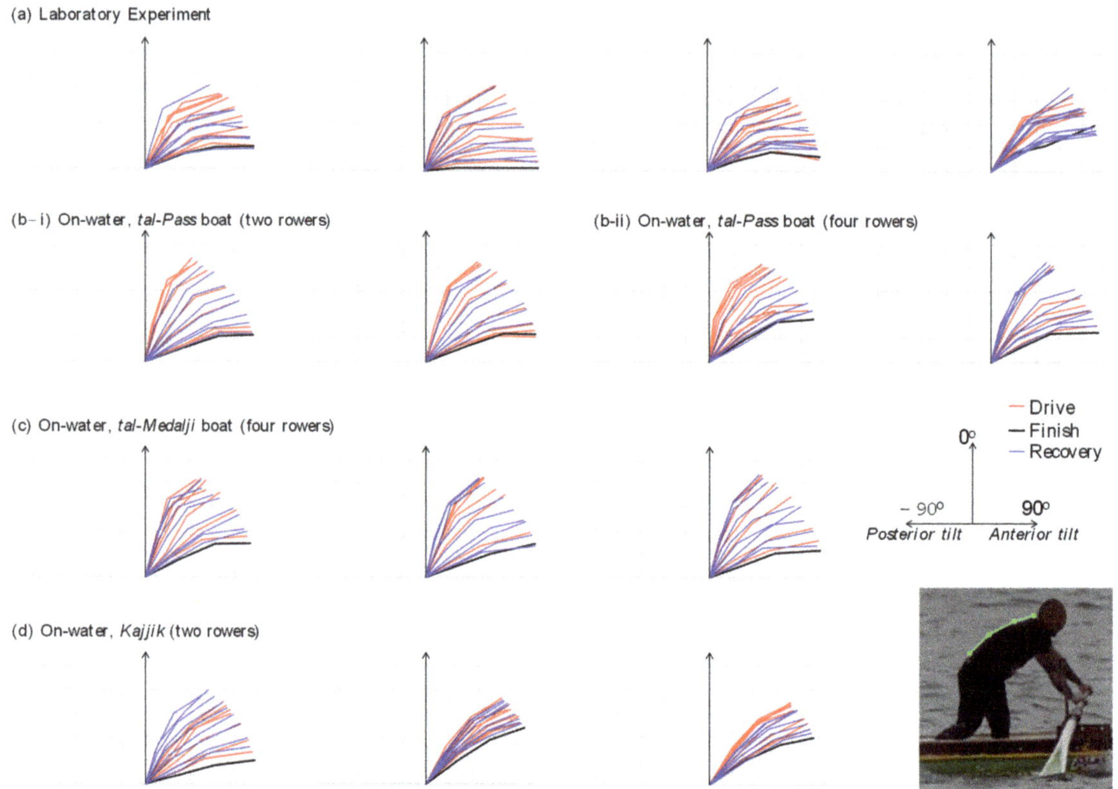

Figure 7. Images showing the typical outline of the back profiles traced from images extracted from video of on-water rowing and the lab study for standing rowing, where (**a**) shows cycles from each of the different rowers in the laboratory study, whilst (**b–d**) refer to the equivalent profiles from rowers rowing on different types of boats. The boats studied are all traditional Maltese rowing racing boats where (**b**) is the widest and heaviest *Tal-Pass* boat which can be rowed by two rowers, one sitting one standing (**b-i**), or four rowers, two sitting, two standing (**b-ii**); (**c**) refers to the more slender, slightly longer, and lighter *Tal-Medalji* version, also rowed by four rowers, two sitting, two standing whilst (**d**) refers to the much shorter *Kajjik*, rowed by two rowers, one sitting, one standing. The rower analysed is always the standing rower with the oar on his right-hand side (*parasija*). The manner of how these profiles were drawn is illustrated in the insert. Note that profiles refer to rowers facing the right direction.

3.3. Measurements concerning the Shoulders and Elbows

Details of the averaged measurement results related to the shoulders and elbow, ± 1.96 standard deviation from finish to finish, are reported in Figures 11 and 12, respectively. For the sagittal plane, these data are also re-plotted, in the form of angle–angle plots, in Figure 13, to highlight the relationships between the various angular measurements and the inclination of the thorax. In addition, reported in Tables 7–9 are the corresponding values for θ_{ROM}, θ_{min} and θ_{max} of each angular measurement plotted in Figures 9 and 10, together with the corresponding p-values related to a comparison between standing rowing and standard ergometer sliding-seat rowing.

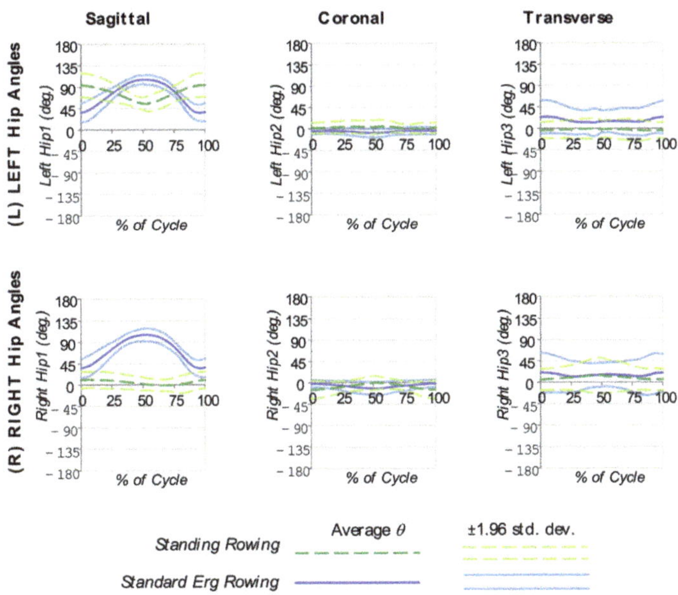

Figure 8. The averaged angular measurements for the hip from the laboratory study, reported as mean ± 1.96 s.d., which compares standing rowing (green) with standard ergometer sliding-seat rowing (blue).

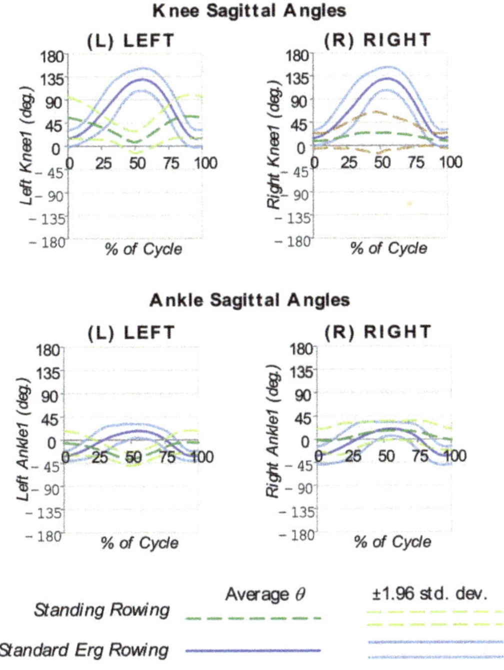

Figure 9. The averaged angular measurements for the knee and ankle (sagittal) from the laboratory study, reported as mean ± 1.96 s.d., which compares standing rowing (green) with standard ergometer sliding-seat rowing (blue).

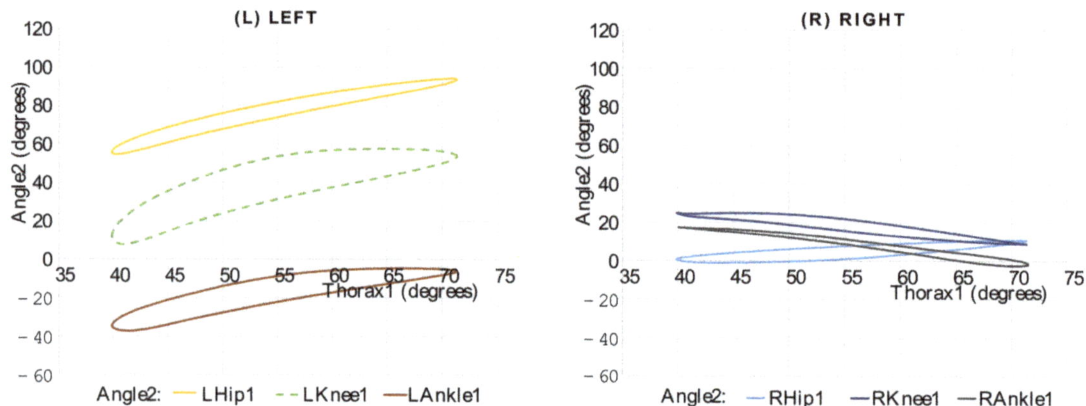

Figure 10. Angle–Angle plots which illustrate the relationship between movements in the lower body with that of the thorax. See also supplementary information for other plots.

Table 4. The range of motion values, θ_{ROM}, of the hip, knee and ankle angles measured for the Venetian style rowing technique, in degrees, $\pm$ 1 s.d., for the individual rowers 1–4, their mean, and its equivalent mean value for standard ergometer rowing, with p-value for these two means reported in parentheses.

θ_{ROM}		Venetian Rowing Technique (VRT)					Control–Erg	
		Rower 1	Rower 2	Rower 3	Rower 4	Mean	Mean	(p)
Hip1	L	43.7 ± 2.5	49.3 ± 4.7	48 ± 1	23 ± 3.6	41 ± 11.4	72.5 ± 9.9	(0.000)
	R	16.3 ± 1.5	23 ± 2.6	6.3 ± 2.1	9 ± 1	13.7 ± 7	72.4 ± 10.2	(0.000)
Hip2	L	8.7 ± 0.6	3.3 ± 0.6	7.3 ± 2.1	5 ± 1	6.1 ± 2.4	7.7 ± 3.4	(0.236)
	R	14 ± 1.7	9.3 ± 2.5	25.7 ± 2.1	12.7 ± 3.2	15.4 ± 6.8	11.2 ± 5.2	(0.162)
Hip3	L	7 ± 1	9.7 ± 1.5	4 ± 1.7	9.7 ± 1.2	7.6 ± 2.7	16.9 ± 6.5	(0.000)
	R	18 ± 8	4.7 ± 1.2	7.3 ± 2.1	17 ± 15.6	11.8 ± 9.7	18.5 ± 5.8	(0.008)
Knee1	L	65.3 ± 1.5	37.7 ± 3.2	64.3 ± 3.1	42.3 ± 4.5	52.4 ± 13.4	116.6 ± 13.1	(0.000)
	R	11.3 ± 2.5	37.7 ± 4.2	30 ± 0	11 ± 3.6	22.5 ± 12.4	116.8 ± 14.6	(0.000)
Ankle1	L	37.7 ± 0.6	28.3 ± 2.5	45 ± 1	26 ± 4.4	34.3 ± 8.2	49.4 ± 5.8	(0.000)
	R	36 ± 3.5	19.7 ± 1.5	14.3 ± 4.2	21.7 ± 5.1	22.9 ± 9	52.7 ± 6.3	(0.000)

Table 5. The minimum values, θ_{min}, of the hip, knee and ankle angles measured for the Venetian style rowing technique, in degrees, $\pm$ 1 s.d., for the individual rowers 1–4, their mean, and its equivalent mean value for standard ergometer rowing, with p-value for these two means reported in parentheses.

θ_{min}		Venetian Rowing Technique (VRT)					Control–Erg	
		Rower 1	Rower 2	Rower 3	Rower 4	Mean	Mean	(p)
Hip1	L	61.7 ± 3.2	42.7 ± 2.5	57.7 ± 1.2	52.7 ± 2.9	53.7 ± 7.7	34.5 ± 8.7	(0.000)
	R	-11.7 ± 1.2	2 ± 1	0.3 ± 0.6	0.3 ± 2.5	-2.3 ± 5.9	33.9 ± 9.2	(0.000)
Hip2	L	7.3 ± 2.1	-3 ± 1	-1.7 ± 1.2	-3.7 ± 3.2	-0.3 ± 5	-9.6 ± 5	(0.000)
	R	-23.3 ± 1.5	-9.7 ± 2.3	-24.3 ± 1.5	-3.7 ± 0.6	-15.3 ± 9.3	-12.5 ± 7.2	(0.346)
Hip3	L	4.3 ± 1.2	-10 ± 1	-18.7 ± 1.5	-1 ± 1	-6.3 ± 9.2	9.8 ± 15.5	(0.004)
	R	10 ± 1	-13.7 ± 1.2	5.3 ± 2.9	16 ± 1.7	4.4 ± 11.7	6.8 ± 17.4	(0.401)

Table 5. *Cont.*

θ_{min}		Venetian Rowing Technique (VRT)					Control–Erg	
		Rower 1	Rower 2	Rower 3	Rower 4	Mean	Mean	(p)
Knee1	L	23 ± 1.7	4.7 ± 0.6	1.3 ± 1.5	-3.3 ± 1.5	6.4 ± 10.5	13.1 ± 8.4	(0.025)
	R	13.7 ± 1.5	13 ± 1	3.7 ± 0.6	-3.7 ± 8.6	6.7 ± 8.4	12.9 ± 7.8	(0.055)
Ankle1	L	-28 ± 2	-43.7 ± 2.3	-41.3 ± 1.2	-40 ± 4.6	-38.3 ± 6.8	-32 ± 4.4	(0.006)
	R	-18.3 ± 5.1	7 ± 1.7	8.7 ± 1.2	-16.7 ± 2.1	-4.8 ± 13.5	-33 ± 9.5	(0.000)

Table 6. The maximum values, θ_{max}, of the hip, knee and ankle angles measured for the Venetian style rowing technique, in degrees, $\pm$ 1 s.d., for the individual rowers 1–4, their mean, and its equivalent mean value for standard ergometer rowing, with p-value for these two means reported in parentheses.

θ_{max}		Venetian Rowing Technique (VRT)					Control–Erg	
		Rower 1	Rower 2	Rower 3	Rower 4	Mean	Mean	(p)
Hip1	L	105.3 ± 1.5	91.7 ± 6.8	105.7 ± 0.6	76 ± 1	94.7 ± 13.1	106.8 ± 4.8	(0.005)
	R	4.7 ± 0.6	25 ± 2.6	6.3 ± 2.1	9.7 ± 3.1	11.4 ± 8.6	106.3 ± 6.9	(0.000)
Hip2	L	16 ± 2	0.7 ± 1.5	5.3 ± 1.2	1 ± 2	5.8 ± 6.6	-2 ± 3	(0.000)
	R	-9 ± 1	-0.3 ± 3.2	1.3 ± 1.2	9.3 ± 2.9	0.3 ± 7.1	-1.4 ± 4.1	(0.000)
Hip3	L	11 ± 2	-0.3 ± 1.5	-15 ± 1	8.7 ± 1.5	1.1 ± 10.7	26.8 ± 16.3	(0.000)
	R	28 ± 8.2	-8.7 ± 0.6	12.3 ± 2.3	33 ± 14	16.2 ± 18.3	25.4 ± 15.9	(0.000)
Knee1	L	88.7 ± 2.9	42 ± 3.6	65.3 ± 1.5	39 ± 5.6	58.8 ± 21.2	129.7 ± 11	(0.000)
	R	24.7 ± 1.5	50 ± 4.6	33.7 ± 0.6	7.3 ± 7.8	28.9 ± 16.6	129.7 ± 11.4	(0.000)
Ankle1	L	9.7 ± 1.5	-15.3 ± 1.5	3.7 ± 0.6	-13.7 ± 3.8	-3.9 ± 11.4	17.5 ± 6.8	(0.000)
	R	17.7 ± 3.5	27 ± 1	23 ± 3.6	6 ± 4.4	18.4 ± 8.7	19.9 ± 7	(0.000)

Table 7. The range of motion values, θ_{ROM}, of the shoulder and elbow angles measured for the Venetian style rowing technique, in degrees, $\pm$ 1 s.d., for the individual rowers 1–4, their mean, and its equivalent mean value for standard ergometer rowing, with p-value for these two means reported in parentheses.

θ_{ROM}		Venetian Rowing Technique (VRT)					Control–Erg	
		Rower 1	Rower 2	Rower 3	Rower 4	Mean	Mean	(p)
Shoulder1	L	64.3 ± 5.9	79.3 ± 2.1	80.3 ± 4.2	84 ± 3.6	77 ± 8.6	101.5 ± 7.9	(0.000)
	R	22.7 ± 3.1	54.7 ± 3.1	35 ± 6.9	85.3 ± 3.1	49.4 ± 25	100.6 ± 9.7	(0.000)
Shoulder2	L	152 ± 3.5	157.7 ± 2.5	162.3 ± 4.7	108 ± 5.3	145 ± 22.9	86 ± 27.7	(0.000)
	R	44.7 ± 7.5	85.7 ± 8.7	75.3 ± 3.2	83.3 ± 11	72.3 ± 18.5	80.2 ± 21.8	(0.591)
Shoulder3	L	137.7 ± 3.8	112.3 ± 4.7	136.3 ± 11.2	116 ± 4.4	125.6 ± 13.3	62.4 ± 28.9	(0.000)
	R	69.3 ± 6	86 ± 12.5	128.7 ± 8.1	90 ± 8	93.5 ± 24	63.6 ± 29.7	(0.014)
Elbow1	L	80 ± 2	74.3 ± 5.9	66 ± 2	66 ± 1.7	71.6 ± 6.8	102.4 ± 6.3	(0.000)
	R	39 ± 5.6	56.3 ± 1.2	33.7 ± 3.2	54.7 ± 3.8	45.9 ± 10.7	98 ± 6.7	(0.000)

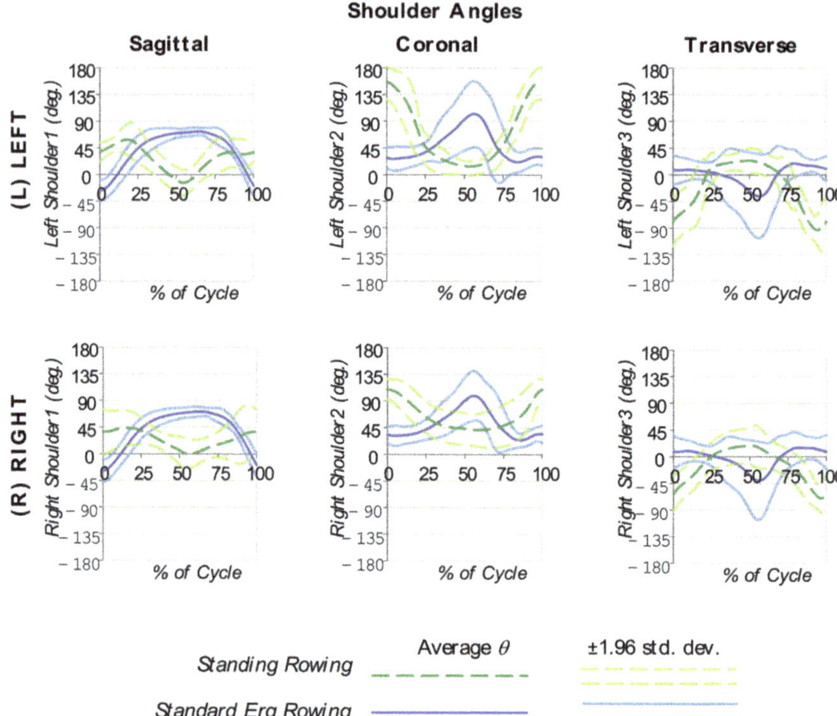

Figure 11. The averaged angular measurements for the shoulders from the laboratory study, reported as mean ± 1.96 s.d., which compares standing rowing (green) with standard ergometer sliding-seat rowing (blue).

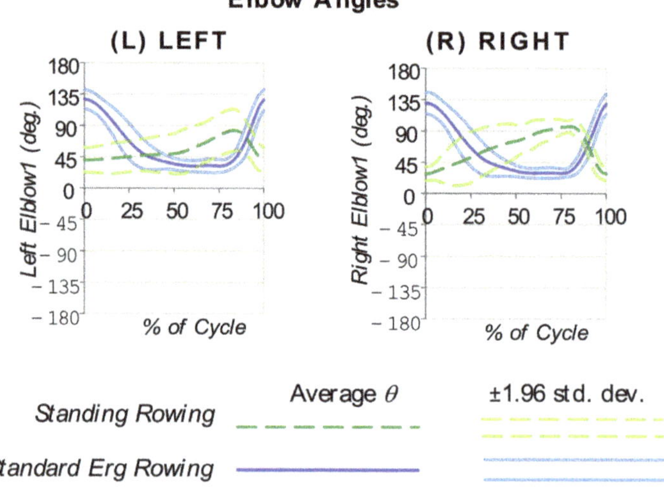

Figure 12. The averaged angular measurements for the elbow (sagittal) from the laboratory study, reported as mean ± 1.96 s.d., which compares standing rowing (green) with standard ergometer sliding-seat rowing (blue).

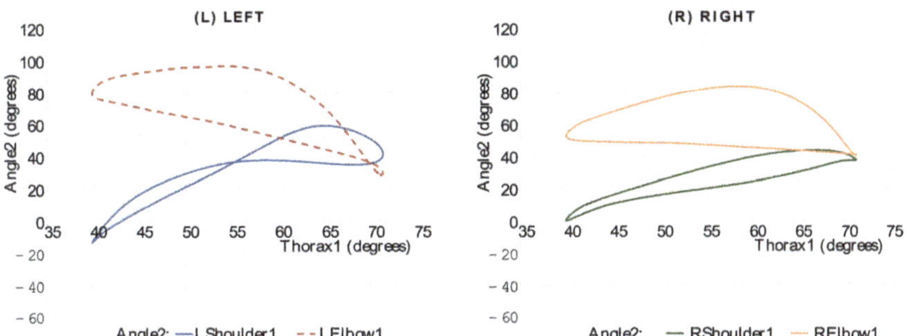

Figure 13. Angle–Angle plots which illustrate the relationship between movements in the shoulders and elbow with that of the thorax. See also Supplementary Materials for other plots.

Table 8. The minimum values, $\theta_{\min}$, of the shoulder and elbow angles measured for the Venetian style rowing technique, in degrees, $\pm$ 1 s.d., for the individual rowers 1–4, their mean, and its equivalent mean value for standard ergometer rowing, with p-value for these two means reported in parentheses.

$\theta_{\min}$		Venetian Rowing Technique (VRT)					Control–Erg	
		Rower 1	Rower 2	Rower 3	Rower 4	Mean	Mean	(p)
Shoulder1	L	-17.3 ± 4.2	-9.3 ± 2.9	-2 ± 2.6	-29 ± 0	-14.4 ± 10.7	-26.7 ± 9	(0.003)
	R	-0.3 ± 1.5	-8 ± 3.5	14 ± 5	-16.3 ± 2.1	-2.7 ± 12	-27.9 ± 9.3	(0.000)
Shoulder2	L	12.3 ± 2.1	6.3 ± 2.1	9.3 ± 2.5	25.3 ± 3.2	13.3 ± 7.9	18.8 ± 7.7	(0.034)
	R	59.3 ± 4.2	27.7 ± 4.9	44.3 ± 1.2	22 ± 3.6	38.3 ± 15.6	21 ± 5.9	(0.000)
Shoulder3	L	-117.3 ± 4.2	-95.7 ± 5.1	-98.3 ± 8.1	-85.3 ± 7.1	-99.2 ± 13.2	-38.6 ± 34.6	(0.000)
	R	-65 ± 1	-76.7 ± 11.6	-83.7 ± 1.5	-64.3 ± 7.5	-72.4 ± 10.4	-42.7 ± 32.6	(0.008)
Elbow1	L	24.3 ± 1.2	24.3 ± 1.5	25.3 ± 0.6	34.7 ± 1.2	27.2 ± 4.6	28 ± 3.6	(0.166)
	R	45.7 ± 2.1	37 ± 3.6	27.3 ± 1.2	42.7 ± 2.1	38.2 ± 7.6	30.5 ± 4.3	(0.008)

Table 9. The maximum values, $\theta_{\max}$, of the shoulder and elbow angles measured for the Venetian style rowing technique, in degrees, $\pm$ 1 s.d., for the individual rowers 1–4, their mean, and its equivalent mean value for standard ergometer rowing, with p-value for these two means reported in parentheses.

$\theta_{\max}$		Venetian Rowing Technique (VRT)					Control–Erg	
		Rower 1	Rower 2	Rower 3	Rower 4	Mean	Mean	(p)
Shoulder1	L	46.7 ± 2.1	70.3 ± 1.2	79 ± 4	55 ± 3.6	62.8 ± 13.5	75 ± 3.4	(0.007)
	R	22.3 ± 1.5	46.7 ± 1.2	49 ± 3.6	69.3 ± 1.5	46.8 ± 17.5	72.7 ± 3.8	(0.007)
Shoulder2	L	164.3 ± 3.5	164 ± 2	171.7 ± 2.3	133 ± 2.6	158.3 ± 15.7	104.6 ± 28.1	(0.000)
	R	104 ± 3.6	113 ± 5.3	119.7 ± 2.1	104.7 ± 9.3	110.3 ± 8.3	101.2 ± 20.5	(0.000)
Shoulder3	L	20.3 ± 1.5	16.7 ± 0.6	38 ± 3.6	31 ± 3.6	26.5 ± 9.1	24 ± 9.3	(0.568)
	R	5 ± 6.6	9.3 ± 4	45 ± 7.2	25.3 ± 0.6	21.2 ± 17	21 ± 11.2	(0.568)
Elbow1	L	104.3 ± 1.2	98.7 ± 7.6	91.3 ± 2.5	100.7 ± 1.5	98.8 ± 6.1	130.3 ± 7.2	(0.000)
	R	84.7 ± 7.4	93.3 ± 4.7	60.7 ± 4	97 ± 3	83.9 ± 15.4	128.5 ± 6.7	(0.000)

4. Discussion

This discussion shall initially focus on the technique of standing rowers in the Venetian style, outlining the similarities to standard rowing and also aspects which are very different.

From a practical aspect, an observation which deserves to be highlighted is that standing rowing is likely to be equally demanding as sitting rowing. Evidence for this claim comes from anecdotal evidence voiced by athletes who practise both forms of rowing and from mechanical analysis of the rigging used in Maltese boats. As noted in Figures 1–3, oarlocks are generally placed in a *quasi*-perfectly anti-symmetric manner, practically equidistantly from the centre of the boat. With such rigging, since the oars used by the standing and sitting rowers are normally equal in length, the boat would only move in a straight direction as long as the standing and sitting rowers put the same amount of power on each respective side. This observation (which needs to be backed up by further studies using protocols typically used on Olympic boats [7,31–36]) is important since, in the absence of specific literature on standing rowing, one can, to a first approximation, hypothesise that findings from studies related to rowing in the standard form (e.g., studies on nutrition, physiology, etc.), are likely to be transferable to athletes who practise standing rowing.

However, sitting and standing rowing differ from each other in a number of other aspects, including very important differences related to movements. In order to expand on the specificities of standing rowing, one would need to highlight what constitutes rowing which, such as walking, running, canoeing and several other sports, can be considered as a cyclic form of movement with numerous repetitions of the same motion. When practised as a sport, in the traditional classical form, the rower sits on a fixed seat and moves the boat primarily by means of the force generated by the trunk and arms with what can be considered as a "pull action" in the "drive" active highly endergonic phase in which the athlete expends energy to move the boat by bringing the oars towards her/himself. This is followed by a much less endergonic "recovery" which sees the athlete reverse back, with the oar outside the water for another active drive phase. This classical form of rowing is normally practised on traditional boats (e.g., the Cornish Pilot gigs, the Italian *Gozzo* or *Lancia*, the Spanish *trainera* or the Maltese "*Dgħajsa tal-Pass*" shown in Figures 1–3), which are wider and heavier than their Olympic counterparts.

With time, the classic form evolved and in Olympic-type rowing, the athlete sits on a mobile seat (an invention which dates to the second part of the 19th century), and consequently the boat is moved with the use of the leg, trunk and arms, with the powerful lower limbs providing most of the power required to propel the boat through a "push action". The "drive" phase in Olympic style rowing initiates at the catch when the blade enters the water with the athlete pushing with their lower limbs whilst keeping the arms straight and the trunk slightly inclined forwards. This is followed by "pulling" actions in the form of tilting the trunk posteriorly (from an "one-o-clock" position to an "eleven-o-clock" position), and finishing with a pull of the arms to bring the oar handles close to the chest and immediately expelling the oar/s out of the water and feather (the "finish" position). Once again, this highly endergonic phase is followed by the "recovery" phase, which essentially sees the athlete reversing from the "finish" position to the "catch" in preparation for the next cycle.

Standing rowing, although practised on traditional boats, differs from the classic fixed-seat seated rowing as it involves a "pushing" action in the endergonic drive phase, rather than pulling. In addition, in contrast with standard sliding-seat rowing, the rower makes extensive use of the upper body in the "pushing action." Such differences are graphically illustrated in Figures 1b and 3 where one can observe the complimentary yet mechanically opposite actions of the seated and standing rowers. These opposing movements are also very evident through the plots of measurements taken in the sagittal plane, i.e., the plane where most movements are recorded, with standing and sitting measurements almost mirroring each other. Thus, for example, the trunk of the standing row would be leaning forwards whilst that of the seated rower is leaning backwards (see Figure 5) or the elbows of the standing rower are in extension whilst those of the seated rower flexed (see Figure 10).

As a result, one could argue that from the perspective of a workout, these two forms of rowing—seating and standing—provide the athlete with a highly complementary set of movements, which if used together could be highly beneficial for the development of the athletes. Moreover, from the perspective of a physical exercise, it is important to highlight that standing rowing, with its demands for stability and balance in the boat (the athlete is standing on a moving and somewhat unstable shell), provides an excellent complimentary exercise to the standard isometric and dynamic balance exercises for the physical conditioning of the athlete.

Keeping these factors in mind, the discussion will now focus on the specific kinematics of standing rowing in the Venetian style, aiming at a better understanding of this technique, enabling replication and possibly further development.

4.1. The Thorax, Pelvis and Spine Kinematics of Standing Rowers

The analysis of the kinematics of standing rowers in the Venetian style is best commenced through the presentation of a selection of consecutive frames within a rowing cycle, such as the ones shown in Figures 3 and 4, focusing on back profiles, which are plotted in Figure 6. These images allow for a very visual appreciation of the main movements associated with this form of rowing, i.e., one where the rower moves from a quasi-standing but inclined forward position at the catch (which corresponds to the sagittal thoracic angle Thorax1 being at θ_{min}, averaging around 40°) to one where the rower, sometimes, has the back parallel to the water surface or the laboratory floor towards the finish (which corresponds to Thorax1 being at θ_{max}, averaged around 72°, but reaching as much as 83° in the case of Rower 2). This latter highly forward flexed position of the thorax is held for quite some time during the rowing cycle and provides two important advantages: (1) It elongates the reach at the finish permitting for a larger arc of travel for the blade that results in forward propulsion (2) the centre of gravity of the rower is lowered, stabilising both himself and the boat. It is interesting to note that the range of motion of the thorax for the standing rowers averages 32°. Here, it must be said that, from the data collected, as well as from actual observation of Venetian style rowing practised on water, the range of motion, as well as extent of forward leaning, seems to be a personal preference of the rowers. Our laboratory study, which looked at four rowers, analysing three cycles per rower, observed one rower who had a range of motion of c. 18°, indicating that the rower performs very little motion forwards and backwards, whilst another had a range of motion of 48° indicating a substantial forward and backwards motion. Irrespective of personal preference, this movement is generally significantly different (lower) to that of the seated rowers, to the extent that the p-value for a comparison between the mean θ_{ROM} for Thorax1 standing vs. standard ergometer rowing is c. 0.

The pelvic movement in the sagittal plane follows a similar trend with the recorded range of motion indicating a synchronised movement between the pelvis and the thorax, with the result that spinal movement is rather low, averaging at 7°. This clearly suggests that the rowers are keeping their lumbar spine erect during the motion with most of the movement occurring at the pelvis (or rather the hip as the pelvis can only attain a small degree of anteversion and retroversion [37].

When viewing the coronal and transverse angular results, it is notable that in contrast to standard ergometer rowing, there is considerable sideways tilting and rotational twisting, with θ_{ROM} averaging at c. 15° for both Thorax2 (coronal) and Thorax3 (transverse). In the case of coronal movements, maximum sideways tilting occurred at the finish. Once again, this indicates that to arrive at the finish with maximum drive length, the trunk exhibits a degree of lateral motion to starboard (in the case of the *Parasija*), with the maximum angle reached by one of the rowers being c. 17° (which was not insignificant). This high range of motion associated with thoracic movements in the coronal plane seems to be a characteristic feature of standing rowing, something which differs significantly from standard sliding-seat erg rowing (p-value < 0.0005). The extent of thoracic transverse plane rotations, Thorax3, is also significantly higher to that observed in standard ergometer rowing, but such rotation is a known feature of sweep rowing as practised by seated rowers. In fact, this rotational component shows that rowers are in a twisted form at the catch and straighten at the finish. This change in posture to assume a non-rotated trunk at the finish has the advantage of a longer and firmer push when reaching the finish point whilst the rather large transverse plane angles at the catch result in an elongation of the drive length.

At this point, it is interesting to make remarks with regard to the individuality shown by the rowers, very evident through the comparison of their thorax. Casual observation of various races and training sessions seems to suggest that (i) different standing rowers adopt different postures and nuances of technique; (ii) rowers may choose to alter their posture during a race, which is generally by design as standing rowers take the role of steering the boat, something which they do by varying slightly their technique to either focus on pushing forward straight, or implementing some directional changes.

4.2. Knee, Ankle and Hip Kinematics of Standing Rowers

One of the most prominent features of standing rowing is that the left and right knees have very asymmetrical movements. In fact, for the *Parasija* position, the rower stands with his feet in a walk-standing posture where the left lower limb leads the movement and moves significantly in the rowing cycle. The range of motion of the left knee averaged at $52°$ (s.d. $13.4°$) with a rather wide range from c. $38°$ to c. $65°$. The catch (which roughly corresponds to the left knee at θ_{min}) and finish angles which roughly correspond to the left knee at θ_{max} were also very individualised, with one rower starting with catch angles as low as c. $-3°$ (i.e., in hyperextension) and other starting at c. $23°$, whilst the finish angles were sometimes as large as c. $89°$ (meaning an almost perpendicular knee stance) to less than $40°$. This limb serves to pivot the body forwards as the rower pushes on the oar. It also provides directional stability during the rowing cycle and involves hip and ankle movements to perform the optimal cycle. This variation may be due to several reasons, such as anthropometric features, and personal preferences such as design, seeking more comfort/stability.

The right knee demonstrates very different behaviour and it seems that some rowers prefer to keep their knee quasi-static throughout the cycle (Rowers 1 and 4) whilst others maintain a flexed position (Rowers 2 and 3). In addition, the lab results suggest that the rowers that do not flex their right knee, tend to assume different positions ranging from a quasi-straight leg during the whole cycle (as was the case with Rower 4) whilst others keep this knee slight bent (as was the case with Rower 1). Nevertheless, in all cases, movements in the left side (for the *Parasija*) are of larger degrees than on the left with visual analysis of on-water rowing confirming that in on-water rowing, some standing rowers tend to flex their left knee to c. $90°$.

Here, it must be re-emphasised that there is a noticeable difference in function between the left and right lower limbs in standing rowing. The right hip serves as the stance leg in that it is posterior to left and pivots mainly on the metatarsal heads of the foot. The right ankle is held in a particular position during standing rowing. Standing rowers all exhibit range of motion changes in the left ankle which is explained by the ankle being in a relative plantar flexed position (around $1°$) at the catch due to the shank being in extension, in line with the rest of the lower limb. This then changes into a dorsiflexed position of around $16.5°$ maximum at the end of the drive phase. Here, it must be noted that the pressure through the ball of the foot through the posterior support plate enhances the overflow of energy and stimulation to contract in the muscles of the right lower limb, a principle known as 'overflow' after Sherrington (1906) [38].

When analysing nuances of the technique of the four standing rowers, it was only Rower 4 who demonstrated barely any movement of the hip, which could be due to his shorter stature and higher BMI and the fact that his technique was remarkably different to the other three rowers. In addition, Rower 1 and Rower 3 were shown to abduct the hip at a certain phase of the cycle whilst Rower 4 maintained lateral rotation throughout the cycle. Once again, through observation, Rower 4 exhibited a rather different movement pattern of the left lower limb, where, besides the laterally rotated hip, the tibia also rocked into lateral rotation and the knee travelled into hyperextension. This pattern was completed with inversion of the foot at the subtalar joint.

Referring to views in the coronal plane, an interesting point was that the taller two of the four rowers (Rower 1 and 3) showed more lateral movement than the other two. This may have the possible implication that taller rowers show more lateral movements; however, one has to keep in mind that techniques vary greatly from one rower to the next.

4.3. The Shoulder and Elbow Kinematics of Standing Rowers

The profile of the angles associated with shoulder movements are unique for standing rowing. For example, a striking feature is that the peaks of the angles in the sagittal plane were not reached at the catch or finish, something that is generally the case with seating rowing, where the catch and finish represent extreme positions. Instead, for the left shoulder, where the arc of shoulder movements is most pronounced for the *Parasija* position, very distinct peak sagittal shoulder angles are encountered twice in cycle: once in mid-recovery (the recovery is initiated with coronal and transverse plane movements which are the most dominant until mid-recovery and the sagittal plane motion only becomes the predominant one after mid recovery), during mid-drive. This double peaking suggests that the shoulder complex plays a very prominent role in standing rowing and the complex and fast movements have to be executed in anatomically and biomechanically correct patterns to produce optimal movements which maximise the length of the drive and its power, in a comfortable yet efficient manner. In the case of the right shoulder, there is some similar, but less pronounced sagittal peaking in mid-recovery, as expected, since both arms must move in synchrony as they hold the same oar. One may observe that peaking in the transverse plane occurs mid-drive, a movement which is necessary due to lateral scapular rotation which allows the glenohumeral joint to rotate superiorly, thus allowing elevation of the humerus with a clear subacromial space. From a performance perspective, this movement at the final part of the drive is essential to elongate the reach at the finish by increasing the arc of the oar blade.

As evidenced from the images depicting standing rowing, there are highly pronounced shoulder movements in standing rowing as the rower uses his shoulder muscles to push, rotate, and stabilise the oar. Once again, apart from discussing the average behaviour, recognising that each rower is different and thus typically performs the rowing action in a somewhat personalised manner, it is also important to discuss any individual variations that one may observe. This personal variation is highly evident in shoulder kinematics, where the data suggest a rather wide range of shoulder angles when comparing the different rowers. However, in general, the angles in the sagittal plane for the left shoulder are at a minimum near the catch. This is explained by the fact that, for standing rowing, the catch commences at a point where the oar is close to the body of the rower as opposed to seated rowing. Thus, the shoulders are kept in a position of extension with the range, once again depending on the anthropometric and rowing-style attributes of the individual rower. At the finish point, the sagittal shoulder angles average to c. 37–38°, with quite a bit of variation between rowers, especially in the left shoulder measurements. This somewhat anomalous value is, when one views the actual action, a dip of the glenohumeral joint towards a slight extension at the finish, which shows the arm in *quasi*-full elevation. However, the fact that the trunk is flexed may explain the interpretation of extension at the finish point. This may possibly be due to the position of the subject in the capture envelope of the motion capture cameras where the angles of the markers are calculated relative to a plane traversing the neck and lower down the pelvis. Hence, with the arm traversing this imaginary plane, the software translates it into an extension movement. However, it must be emphasised that maximum shoulder angles are not demonstrated neither at the catch nor at the finish, but halfway through the recovery as the rower is pushing back the oar to its original catch position. It must also be said that the profile of the blade in the water for standing rowing is much more circular than oval (as is the case of sitting rowing), thus explaining the wide range of motion of the shoulder angles. The coronal movement also reveals differences between the left and right shoulder angles where the left goes through a larger arc of abduction compared to the right.

In the case of the elbow movements for standing rowing, once again, the most interesting feature is that the peak movement is occurring mid-drive and not at the catch or finish. This movement corresponds to the final push that rowers give to the oars just prior to the finish, which in the case of seated rowers is accomplished by feathering in the water, and which gives the heavy boats the impulse required to keep their speed during the recovery.

4.4. Strengths and Limitations of This Work

This work investigated the biomechanics and kinematics of Venetian style standing rowing and compares it with what is known about the standard and traditional seated forms of rowing. The main strength of this work is that this work reports the kinematics of a technique that has never, to the authors' knowledge, been studied quantitatively and documented for other to replicate and further develop. In doing so, this work identified some aspects that were either previously ignored or never formally recorded or studied. These include the fact that for the first time, it is acknowledged that Venetian-type rowing may be as efficient and effective as its seated counterpart, and in many aspects complementary to it. All this will hopefully permit better recognition of this most elegant rowing technique, and possibly also encourage others to try it. From a very practical aspect, it should be noted that being forward-facing, standing rowing could be particularly useful to help one navigate and steer a boat in busy stretches of water.

An important strength is that methodology adopted, which examines the rowers in a state-of-the-art calibrated laboratory setting rather than on-water, eliminates the limitations associated with taking on-water measurements. However, the use of on-water data to compliment the laboratory findings gives peace of mind that what is being reported here is likely to be a true replica of the real on-water scenario. Moreover, this work has shown how a standard sliding-seat rowing ergometer can be further developed to emulate standing rowing, something which is likely to be appealing to athletes, coaches and gym users as a novel and effective training tool for physical conditioning. Recognising that the sports industry is known to be constantly on the lookout for new ideas for developing new tools and devices for physical conditioning and a complete workout that replicates the sports movement (e.g., rowing ergometer, canoe ergometer, the ski-erg, etc.), it is hoped that this concept will be further developed and ultimately become mainstream.

Limitations include the fact that with work focused purely on kinematic aspects in a highly idealistic laboratory scenario which, for example, did not have into account the instability provided by the water, which might influence motor control and kinematic parameters. Another limitation of this work was that the rowers were permitted to row at their own pace. Whilst this had the advantage that rowers felt at ease during the experiment, it had the disadvantage that the data generated needed to be processed in order to make the catch and finish points coincide (a simple process where the average occurrence of the "catch" within the "finish to finish" cycles was identified and the length of the drive and recovery phases appropriately scaled so that "finish" and "catch" are synchronised). Moreover, the number of subjects studied was small and no attempt was made to standardise the data (apart from reporting the cycle as a percentage and aligning the 'catch' and 'finish'). Given that it is well known that kinematics is highly affected by height, span, etc., it would be useful if such factors are taken into consideration in further studies. Moreover, it would have been ideal to link technique to results obtained in races, something which was not performed for ethical reasons. Moreover, no attempt was made to study how standing rowing could physically affect the athletes practising it (e.g., changes in body composition, effect on strength, muscle mass, tendons, bone strength, etc. [39,40]), or how athletes alter their technique when they have muscle tightness [30], become fatigued or blistered [26], or if they need to augment or decrease the power to increase boat speed or to steer the boat. Ideally this study should be extended to look into these issues, measure muscle activity, forces, etc.

Another important limitation is that this work only looked at rowers who hold the oar on their left-hand side (the "parasija" position). Whilst there are no reasons to suggest that the standing rowers on the other board would row differently, apart from rowing in "mirror image", it would be useful if in the future the kinematics of rowers on the opposite board are also studied.

Finally, an important limitation is that this work only looked at Venetian-style standing rowing through the manner how this is practised as a sport in Malta. Whilst this had the advantage that in Malta, standing rowers row on the same boat with the same stroke as sitting rowers, it would have been useful to also examine rowers who hail from different countries to assess if the technique practised in Malta differs in any way from that practised in other countries.

5. Conclusions

This work has for the first time,

- Documented the kinematics of standing Venetian style rowing, as performed by abled-bodied athletes through a laboratory-based study supplemented by data recorded on-water in a manner which permits this historical form of rowing made famous by the gondoliers of Venice to be replicated by others and form the basis for further studies and developments;
- Measured and reported the range of motion of various joints, thus showing that it constitutes a whole-body exercise which is highly complementary to standard and traditional seated rowing;
- Highlighted some unique advantages associated with this form of rowing, such as that it can be performed by an athlete facing forward rather than backwards, thus possibly making it safer to practise in busy waters.

Hopefully, with this first ever toolkit to help athletes and coaches clearly understand how competitive Venetian-style standing rowing is practised, this form of rowing will grow in popularity through a wider uptake by new athletes and further studies. In particular, further studies could look into muscle activity, energetics, injuries, etc., and also look into the possibility of making this form of rowing accessible to individuals with impairments, in an analogous way that para-rowing exists alongside Olympic rowing, something which so far seems to have never been attempted for Venetian-style rowing.

Supplementary Materials: The following supporting information can be downloaded at: https://www.mdpi.com/article/10.3390/bioengineering10030310/s1, Figure S1: Angle-Angle graphs, plotted against Thorax1; Figure S2: Angle-Angle graphs, plotted against LKnee1; Figure S3: Angle-Angle graphs, plotted against LElbow1. Table S1: Detailed statistical information related to angular measurements (in degrees).

Author Contributions: Conceptualization, J.N.G. and T.P.A.; Data curation, D.S. and T.P.A.; Formal analysis, J.N.G. and A.S.; Funding acquisition, J.N.G., A.S. and T.P.A.; Investigation, J.N.G. and T.P.A.; Project administration, J.N.G. and T.P.A.; Resources, C.F. and A.G.; Software, D.S., M.G. and R.N.; Supervision, J.N.G. and J.X.d.C.; Validation, A.S. and T.P.A.; Visualization, J.N.G.; Writing—original draft, J.N.G., D.C. and T.P.A.; Writing—review and editing, J.N.G., D.C., N.C. and T.P.A. All authors have read and agreed to the published version of the manuscript.

Funding: J.N.G. acknowledges the funding received from the Malta Council for Science & Technology (A-ROW, FUSION: The R&I Technology Development Programme 2018 project), grant number R&I-2017-033T. J.N.G., T.A. and A.S. acknowledge the support of research grants awarded by the University of Malta.

Institutional Review Board Statement: The study was conducted according to the guidelines of the Declaration of Helsinki, and approved by the University Research Ethics Committee [UREC], of The Faculty of Health Sciences, University of Malta (FREC FHS_1718_017) on the 28 March 2018.

Informed Consent Statement: Informed consent was obtained from all subjects involved in the study.

Data Availability Statement: Data are contained within the article.

Acknowledgments: The authors acknowledge the support received by the various athletes who participated in this study. The contribution of James English (Staffordshire University, UK), Juan Farrugia and Jan Tanti, who helped in constructing the setup, and that of Thomas Camilleri Mallia, Henrietta Camilleri Mallia and Michele Agius, who provided the video and photography footage, is most gratefully acknowledged. J.N.G. is particularly grateful to Aaron R Casha (1966–2020), his friend and colleague, who shared many hours of his life discussing biomechanics, rowing, and many other topics of mutual interest.

Conflicts of Interest: The authors declare no conflict of interest. The funders had no role in the design of the study, in the collection, analyses, or interpretation of data, in the writing of the manuscript, or in the decision to publish the results.

References

1. Capelli, C.; Tarperi, C.; Schena, F.; Cevese, A. Energy cost and efficiency of Venetian rowing on a traditional, flat hull boat (Bissa). *Eur. J. Appl. Physiol.* **2009**, *105*, 653–661. [CrossRef] [PubMed]
2. Capelli, C.; Donatelli, C.; Moia, C.; Valier, C.; Rosa, G.; di Prampero, P.E. Energy cost and efficiency of sculling a Venetian gondola. *Eur. J. Appl. Physiol. Occup. Physiol.* **1990**, *60*, 175–178. [CrossRef] [PubMed]
3. Nolte, V. (Ed.) *Rowing Faster*, 2nd ed.; Human Kinetics, Inc.: Champaign, IL, USA, 2011; ISBN 9780736090407.
4. Černe, T.; Kamnik, R.; Vesnicer, B.; Žganec Gros, J.; Munih, M. Differences between elite, junior and non-rowers in kinematic and kinetic parameters during ergometer rowing. *Hum. Mov. Sci.* **2013**, *32*, 691–707. [CrossRef] [PubMed]
5. Kleshnev, V. Biomechanics for Rowing Technique and Rigging. In Proceedings of the World Rowing Youth Coaches Conference, Tel Aviv, Israel, 14–17 October 2010.
6. Wilson, F.; Gissane, C.; Gormley, J.; Simms, C. Sagittal plane motion of the lumbar spine during ergometer and single scull rowing. *Sport. Biomech.* **2013**, *12*, 132–142. [CrossRef]
7. Hume, P.A. Movement analysis of scull and oar rowing. *Handb. Hum. Motion* **2018**, *2–3*, 1719–1739. [CrossRef]
8. Soper, C.; Hume, P.A. Towards an ideal rowing technique for performance: The contributions from biomechanics. *Sport. Med.* **2004**, *34*, 825–848. [CrossRef]
9. Marcolin, G.; Lentola, A.; Paoli, A.; Petrone, N. Rowing on a boat versus rowing on an ergo-meter: A biomechanical and electromyographycal preliminary study. *Procedia Eng.* **2015**, *112*, 461–466. [CrossRef]
10. Shephard, R.J. Science and medicine of rowing: A review Science and medicine of rowing: A review. *J. Sports Sci.* **1998**, *16*, 603–620. [CrossRef]
11. Mahler, D.A.; Nelson, W.N.; Hagerman, F.C. Mechanical and Physiological Evaluation of Exercise Performance in Elite National Rowers. *JAMA J. Am. Med. Assoc.* **1984**, *252*, 496–499. [CrossRef]
12. Cutler, B.; Eger, T.; Merritt, T.; Godwin, A. Comparing para-rowing set-ups on an ergometer using kinematic movement patterns of able-bodied rowers. *J. Sports Sci.* **2017**, *35*, 777–783. [CrossRef]
13. Cutler, B.; Merritt, T.; Eger, T.; Godwin, A. Using Peak Vicon data to drive Classic JACK animation for the comparison of low back loads experienced during para-rowing. *Int. J. Hum. Factors Model. Simul.* **2015**, *5*, 99. [CrossRef]
14. Halliday, S.E.; Zavatsky, A.B.; Andrews, B.J.; Hase, K. Kinematics of the Upper and Lower Extremities in Three-Dimensions during Ergometer Rowing. In Proceedings of the 18th Congress of the International Society of Biomechanics, Zurich, Switzerland, 8–13 July 2001; pp. 22–24.
15. Treff, G.; Leppich, R.; Winkert, K.; Steinacker, J.M.; Mayer, B.; Sperlich, B. The integration of training and off-training activities substantially alters training volume and load analysis in elite rowers. *Sci. Rep.* **2021**, *11*, 17218. [CrossRef]
16. Volianitis, S.; Yoshiga, C.C.; Secher, N.H. The physiology of rowing with perspective on training and health. *Eur. J. Appl. Physiol.* **2020**, *120*, 1943–1963. [CrossRef]
17. Cerasola, D.; Bellafiore, M.; Cataldo, A.; Zangla, D.; Bianco, A.; Proia, P.; Traina, M.; Palma, A.; Capranica, L. Predicting the 2000-m Rowing Ergometer Performance from Anthropometric, Maximal Oxygen Uptake and 60-s Mean Power Variables in National Level Young Rowers. *J. Hum. Kinet.* **2020**, *75*, 77–83. [CrossRef]
18. Cataldo, A.; Cerasola, D.; Russo, G.; Zangla, D.; Traina, M. Mean power during 20 sec all-out test to predict 2000 m rowing ergometer performance in national level young rowers. *J. Sport. Med. Phys. Fit.* **2015**, *55*, 872–877.
19. Cerasola, D.; Cataldo, A.; Bellafiore, M.; Traina, M.; Palma, A.; Bianco, A.; Capranica, L. Race profiles of rowers during the 2014 Youth Olympic Games. *J. Strength Cond. Res.* **2018**, *132*, 2055–2060. [CrossRef]
20. Cerasola, D.; Zangla, D.; Grima, J.N.; Bellafiore, M.; Cataldo, A.; Traina, M.; Capranica, L.; Maksimovic, N.; Drid, P.; Bianco, A. Can the 20 and 60 s All-Out Test Predict the 2000 m Indoor Rowing Performance in Athletes? *Front. Physiol.* **2022**, *13*, 776. [CrossRef]
21. Cerasola, D.; Cataldo, A.; Bianco, A.; Zangla, D.; Capranica, L.; Traina, M. Drag Factor on Rowing Ergometer During 2000-M Performance in Young Rowers. *Kinesiol. Slov.* **2017**, *23*, 15–21.

22. Bernardes, F.; Mendes-Castro, A.; Ramos, J.; Costa, O. Musculoskeletal Injuries in Competitive Rowers Lesões Músculo-Esqueléticas em Remadores de Competição. *Acta Med. Port.* **2015**, *28*, 427–434. [CrossRef]
23. Rumball, J.S.; Lebrun, C.M.; Di Ciacca, S.R.; Orlando, K. Rowing injuries. *Sport. Med.* **2005**, *35*, 537–555. [CrossRef]
24. Wilson, F.; Gissane, C.; Gormley, J.; Simms, C. A 12-month prospective cohort study of injury in international rowers. *Br. J. Sports Med.* **2010**, *44*, 207–214. [CrossRef] [PubMed]
25. Karlson, K.A. Rowing injuries. *Phys. Sport. Med.* **2000**, *28*, 40–50. [CrossRef] [PubMed]
26. Grima, J.N.; Wood, M.V.; Portelli, N.; Grima-Cornish, J.N.; Attard, D.; Gatt, A.; Formosa, C.; Cerasola, D. Blisters and Calluses from Rowing: Prevalence, Perceptions and Pain Tolerance. *Medicina* **2022**, *58*, 77. [CrossRef] [PubMed]
27. Lvov, V.A.; Senatov, F.S.; Veveris, A.A.; Skrybykina, V.A.; Díaz Lantada, A. Auxetic Metamaterials for Biomedical Devices: Current Situation, Main Challenges, and Research Trends. *Materials* **2022**, *15*, 1439. [CrossRef]
28. Arumugam, S.; Ayyadurai, P.; Perumal, S.; Janani, G.; Dhillon, S.; Thiagarajan, K.A. Rowing Injuries in Elite Athletes: A Review of Incidence with Risk Factors and the Role of Biomechanics in Its Management. *Indian J. Orthop.* **2020**, *54*, 246–255. [CrossRef]
29. Ozturk, A.; Tartar, A.; Ersoz Huseyinsinoglu, B.; Ertas, A.H. A clinically feasible kinematic assessment method of upper extremity motor function impairment after stroke. *Meas. J. Int. Meas. Confed.* **2016**, *80*, 207–216. [CrossRef]
30. You, J.Y.; Lee, H.M.; Luo, H.J.; Leu, C.C.; Cheng, P.G.; Wu, S.K. Gastrocnemius tightness on joint angle and work of lower extremity during gait. *Clin. Biomech.* **2009**, *24*, 744–750. [CrossRef]
31. Kleshnev, V. Comparison of measurements of the force at the oar handle and at the gate or pin. *Rowing Biomech. Newsl.* **2010**, *10*.
32. Baudouin, A.; Hawkins, D. Investigation of biomechanical factors affecting rowing performance. *J. Biomech.* **2004**, *37*, 969–976. [CrossRef]
33. Warmenhoven, J.; Cobley, S.; Draper, C.; Smith, R. Over 50 Years of Researching Force Profiles in Rowing: What Do We Know? *Sport. Med.* **2018**, *48*, 2703–2714. [CrossRef]
34. Coker, J.; Hume, P.; Nolte, V. Validity of the Powerline Boat Instrumentation System. In Proceedings of the 27th International Conference on Biomechanics in Sports, Limerick, Ireland, 17–21 August 2009; pp. 65–68.
35. Lukashevich, D.; Huseynov, D.; Minchenya, A.; Bubulis, A.; Vėžys, J. Smart sensors for estimation of power interaction of an athlete with water surface when paddling in the cycle of rowing locomotions. *J. Complex. Health Sci.* **2020**, *3*, 81–90. [CrossRef]
36. Smith, T.B.; Hopkins, W.G. Measures of rowing performance. *Sport. Med.* **2012**, *42*, 343–358. [CrossRef]
37. Norkin, C.C.; White, D.J. *Measurement of Joint Motion: A Guide to Goniometry*, 5th ed.; F. A. Davies: Philadelphia, PA, USA, 1986; Volume 40, ISBN 9780803645660.
38. Sherrington, C.S. *The Integrative Action of the Nervous System*; Yale University Press: London, UK, 1906.
39. Gatt, R.; Vella Wood, M.; Gatt, A.; Zarb, F.; Formosa, C.; Azzopardi, K.M.; Casha, A.; Agius, T.P.; Schembri-Wismayer, P.; Attard, L.; et al. Negative Poisson's ratios in tendons: An unexpected mechanical response. *Acta Biomater.* **2015**, *24*, 201–208. [CrossRef]
40. Ertas, A.H.; Winwood, K.; Zioupos, P.; Cotton, J.R. Simulation of creep in non-homogenous samples of human cortical bone. *Comput. Methods Biomech. Biomed. Engin.* **2012**, *15*, 1121–1128. [CrossRef]

Article

Influence of Different Load Conditions on Lower Extremity Biomechanics during the Lunge Squat in Novice Men

Lidong Gao [1], Zhenghui Lu [1], Minjun Liang [1,2,*], Julien S. Baker [3] and Yaodong Gu [1,4,*]

1 Faculty of Sports Science, Ningbo University, Ningbo 315211, China; gaolidong1997@hotmail.com (L.G.); luzhenghui_nbu@foxmail.com (Z.L.)
2 Department of Physical and Health Education, Udon Thani Rajabhat University, Udon Thani 41000, Thailand
3 Center for Health and Exercise Science Research, Department of Sport and Physical Education, Hong Kong Baptist University, Hong Kong; jsbaker@hkbu.edu.hk
4 Savaria Institute of Technology, Eötvös Loránd University, H-9700 Szombathely, Hungary
* Correspondence: liangminjun@nbu.edu.cn (M.L.); guyaodong@nbu.edu.cn (Y.G.)

Abstract: Objective: The lunge squat is one of the exercises to strengthen the lower limbs, however, there is little evidence of the effects of different equipment. The purpose of this study was to investigate the biomechanical effects of different types of equipment and loads on the lunge squat's effect on the lower limbs. Methods: Fourteen male fitness novices participated in the experiment. Kinematics and kinetics in the sagittal plane using dumbbells, barbells, and weighted vests were measured using OpenSim. Two-way repeated measures ANOVA and one-dimensional statistical parametric mapping were used in the statistical analysis (SPM1D). Results: Range of motion (ROM) change in the knee joint was more obvious when using a barbell, whereas ROM when using a dumbbell was minimal. Compared to other joints, the joint moment at the hip joint was the largest and changed more significantly with increasing weight-bearing intensity, and the change was more pronounced with the dumbbell. For the center of pressure (COP) overall displacement, the dumbbell produced a smaller range of displacement. Conclusions: Dumbbells are suggested for male beginners to improve stability, barbells for the more experienced, and a low-weighted vest may be more appropriate for those with knee pain.

Keywords: lunge squat; kinematic; kinetics; weight bearing; OpenSim

Citation: Gao, L.; Lu, Z.; Liang, M.; Baker, J.S.; Gu, Y. Influence of Different Load Conditions on Lower Extremity Biomechanics during the Lunge Squat in Novice Men. *Bioengineering* **2022**, *9*, 272. https://doi.org/10.3390/bioengineering9070272

Academic Editors: Rui Zhang and Wei-Hsun Tai

Received: 8 May 2022
Accepted: 20 June 2022
Published: 22 June 2022

Publisher's Note: MDPI stays neutral with regard to jurisdictional claims in published maps and institutional affiliations.

1. Introduction

The lunge squat is becoming increasingly popular as a strength training exercise because of its safety and feasibility [1,2]. Performing the lunge squat requires both feet to be placed on the ground. This can also be a suitable exercise for people with poor lower limb balance [3]. The lunge squat is also a common position used in many sports, and a suitable lunge training program is important to improve athletic performance and rehabilitation after surgery [4].

As a closed-chain exercise modality [4–7], the lunge squat is performed by mobilizing multiple joints in a coordinated movement. It improves body coordination and has a positive effect on the training of the lower limb and hip muscles [8]. The lunge squat also plays a key role in rehabilitation exercises used for static and dynamic joint balance [1,8]. Therefore, the action is often used in clinical settings, such as rehabilitation following anterior cruciate ligament (ACL) reconstruction surgery [4,9] and as an exercise prescription for fall prevention exercises [2]. Furthermore, because of the strengths and weaknesses in these areas that have been associated with injury development [10–13], postoperative recovery of the hip and knee joints should be given particular priority. The strength and endurance of the hip muscles, on the other hand, are critical for injury prevention, gait correction, pain alleviation, and improved athletic performance [14–17]. For example, strengthening the gluteus medius has been proven to be helpful in functional recovery and pain reduction in patients after knee meniscus surgery [15]. In addition, maintaining the gluteus maximus muscle helps reduce the incidence of low back pain and disability [16]. Because of the better training effect on the large muscle groups of the lower limbs, the lunge squat is an exercise that requires physical strength and muscular endurance [18]. It is worth noting that the lunge squat effectively increases the strength of the hip, knee, and ankle joints, as well as the quadriceps [8,19]. In addition, increasing the muscle activity ratio between the medial femoral and lateral femoral muscles through effective exercise maintains strength balance between the right and left muscles around the knee joint [8,20]. A study demonstrated that the lunge effectively improves muscle strength and balance in the lower limbs [3]. The hip muscles' activation degree significantly affects the ability of the quadriceps and leg muscles to generate force or resist impact during jumping [21]. Additionally, improved hip muscle function may help to prevent common lower limb problems such as anterior cruciate ligament injuries [6]. Farrokhi et al. [22] found that when performing a forward lunge with a more forward-leaning trunk, the extensor impulse and electromyography of the hip joint were significantly increased, PFA was greater than in upright lunges, and hip extensors were enhanced more in forward trunk lunges.

Adding greater weights is the most popular and important method of resistance training for athletes to improve athletic performance [23]. Compared to the ankle and hip joints, it was found that the lunge squat relies more on the knee joints. In addition, the increased external weight causes more mechanical work to be performed at the hip and ankle joints [9]. At the same time, differences in body posture and changes in stride length can affect the biomechanical changes of each joint during the lunge squat. Previous studies have found that increasing the dominant leg's tibia angle to the ground resulted in a smaller range of motion (ROM) in the anterior knee and a larger ROM in the posterior knee and hip. Meanwhile, as the tibial angle increases, the flexion moment of the anterior leg knee and anterior leg hip decreases, but the flexion moment of the anterior hip joint increases as stride length increases [24]. It has also been observed that muscle activation was significantly higher in the weight-bearing condition than in the self-weight condition when only using barbells, dumbbells, weighted vests, and kettlebells. Quadriceps and hamstring activity were significantly higher in the lunge-step exercise than in the deep squat. However, Wu et al. [4] argued that muscle activation was not affected by changes in weight-bearing equipment during lunge-stepping, but in their study, the subjects were active and only the muscle activation level was studied. Other aspects of biomechanics were not addressed, such as maximum peak angle and peak moment. Choosing the right type of load condition improves the exerciser's results [19]. In addition, compared to the deep squat, the lunge squat requires more balance due to the movement of the center of gravity on the ground. The external force applied to the body also follows the changes in force, causing the mobilization of many muscles to maintain body stability [8]. Experienced fitness enthusiasts demonstrate greater muscular strength and better joint coordination, producing higher power and utilizing muscle force generation through joint coordination [25]. Whereas for fitness novices, muscle imbalance and poor body coordination, combined with a tendency to lean forward during the lunge, increases the shear force at the knee joint, increasing the risk of injury [26,27]. Fitness instructors should develop the most suitable lunge exercise program for fitness participants based on experience.

A large number of studies have focused on different exercise types, such as comparing biomechanical changes and muscle activation in the deep squat, single-leg squat, and lunge squat [4,5,8,10,18,24,28–30]. Some studies have focused on the effects of different postures, such as the degree of inclination of the dominant calf and lunge step length [22,24,31,32]. All of these studies used load weight fixing in these tests. In contrast, Bryan [9] and Nadza-lan [7] did a longitudinal study using weights, comparing changes in the hip, knee, and ankle at different load intensities while focusing on young people. With the improvement of young people's living standards and health awareness, an increasing number of males are concerned about their physical development [33] and they want to develop their muscles using scientific perspectives, relying not only on subjective experience in selecting exercise movements. However, novice athletes are inefficient and at risk of injury due to a lack of theoretical knowledge and irrational movement choices [34]. Therefore, the main purpose of this study was to investigate the kinematic and kinetic changes of the hip, knee, and ankle joints of male fitness beginners at 25% body weight (BW) and 50% BW with dumbbells, barbells, and weighted vests using OpenSim, and to provide appropriate guidance based on the results of the study. Therefore, three hypotheses are presented: (a) The three weight types have different degrees of influence on the kinematics and kinetics of the hip, knee, and ankle. (b) The influence of the weight types on the joint deepens as the intensity of the weight increases. (c) There are one or more appropriate weight types and intensity combinations due to the differences in the exercise effect of the different weight types.

2. Materials and Methods

2.1. Participants

Fourteen male fitness novices were recruited for this study (age: 22.7 ± 1.6 years, height: 1.72 ± 0.05 m, weight: 73.17 ± 0.02 kg, BMI: 24.89 ± 1.52 kg/m^2, leg length: 0.86 ± 0.04 m, shoulder width: 0.41 ± 0.03 m). The dominant leg was identified by preference during kicking and the preferred limb was defined as the dominant limb. There were no lower limb diseases or injuries in the six months before testing. Participants did not eat for 2 h before the experiment and were prohibited from consuming any type of alcohol or caffeine for 24 h. They all wore the same style of shoes to avoid the effects of shoes. All participants understood the purpose and significance of the study, were informed of the testing procedures, and signed an informed consent form before experimental data collection. The Ethics Committee of Ningbo University Research Institute approved the experiment (No: RAGH202110283004.2).

2.2. Instruments

An eight-camera Vicon motion capture system (Vicon Metrics Ltd., Oxford, UK) was used to capture motion trajectories. An embedded Kistler 3D force plate (Kistler, Switzerland) was used to record ground reaction forces simultaneously at 200 Hz and 1000 Hz. Before the experiment, subjects were asked to apply 41 reflective markers (diameter: 14 mm) to their bodies. The specific location points are outlined in Figure 1. The midpoint of the femur was defined as the central point of the greater trochanter and the epicondyle of the femur. The midpoint of the fibula was defined as the central point of the lateral condyle and the lateral ankle.

Figure 1. (**a**): Front, side, and back view of the subject's reflective markers; (**b**): vest lunge pose; (**c**): barbell start pose; (**d**): dumbbell lunge pose; (**e**): lunge start pose.

2.3. Procedures

Participants performed a 5 min warm-up on a stationary treadmill before performing a lunge squat. The leg length of each participant was measured using a metric ruler. Leg length was defined as the length from the anterior superior iliac spine to the medial tibial condyle. We defined stride length as 70% leg length and marked the corresponding stride length with tape at the start point and target stride point (on the force platform) as the lunge stride length for each participant.

Before the experiment began, participants underwent practice experiments for each weight type; subjects were instructed by related professional practitioners, and verbal instruction ensured familiarity with the movements without major problems. At the beginning of the experiment, the participants stood with their feet together at the starting point, and after hearing the command, they moved the dominant leg forward to the marked position. They then shifted their body weight in a downward direction and kept their trunk upright. In the end, the non-dominant leg moved forward and landed in a position parallel to the dominant leg. The body weight returned to the initial state and ended with the body upright and knees fully extended as the end of a full movement. We specified that a qualified movement required: (1) the torso to remain upright and perpendicular to horizontal, with the arms always on either side of the torso and without swaying; (2) the knee joint of the dominant leg to go no further forward than the toe and the lower leg perpendicular to the ground; (3) the knee joint of the non-dominant leg to be close to the ground but not touching the ground. When all three conditions were met, it was deemed a qualified action.

Participants performed a total of six sets of experiments, with each individual's weight-bearing intensity and order of weight-bearing type determined by completely random selection. Six qualifying data were collected for each condition, from which three lunges were selected. After each set was completed, participants rested for 5 min to prevent fatigue. The lunge squat had six weighting situations with 25% and 50% BW of dumbbells, barbells, and weighted vests [9,24], as outlined in Figure 1. The barbell was kept in position behind the back and the dumbbells of the same weight were held by both the left and right upper limbs. Vest contact with the subject's skin ensured that it was always fixed.

2.4. Data Collection and Processing

One lunge data period was defined as starting from the previous frame of the dominant leg contacting the force table to the end of the next stance position. The kinematic and kinetic data (including peak flexion angle (PFA), ROM, and peak moments, as well as center of pressure (COP)) of the hip, knee, and ankle joints in the sagittal planes were collected to analyze joint changes under different load conditions.

Each set of data for PFA, ROM, peak moment, and COP was normalized to 0–100%, whereas subjects' body weight and weight after weight bearing were used to normalize joint moments. COP was normalized for leg length to eliminate the effect of height [35].

Marker trajectories and ground reaction forces were filtered by zero-latency fourth-order Butterworth low-pass filters at 12 Hz and 30 Hz, respectively. A threshold of 20 N was used, and ground reaction force data below 20 were rejected. The C3D file data were converted to formats recognized in OpenSim 4.3 (.mot and .trc) by Matlab R2018a (The MathWorks, MA, USA), and then imported into OpenSim for data processing [36]. The model employed in this experiment was adapted by Lu et al. [37] based on the open-source original model of the deep squat [38] to prevent muscles from crossing bones and causing higher joint mobility. The model was scaled using the subject's marker point location and weight in a static calibration. The static weight of each marker was manually adjusted according to the root mean square (RMS) error value (less than 0.02) between the experimental and virtual markers in the model until it was adjusted to the appropriate position before applying the scaled model to the data calculation. The joint angles were calculated using the inverse kinematics (IK) calculation tool in OpenSim and the results were optimized using least squares to minimize the error between the experimental and virtual markers. The inverse dynamics (ID) algorithm was used to calculate the net moments of the hip, knee, and ankle joints. The ID tool performs inverse dynamic analysis by applying these data given the kinematics describing the movement of the model and perhaps a portion of the kinetics applied to the model. The ID tool solves the mathematical equations of force and acceleration of classical mechanics in an inverse dynamics sense to yield the net forces and torques at each joint which produce the movement [39].

2.5. Statistical Analysis

In this study, statistical parametric mapping was used to compare joint angles and moments, and one-dimensional statistical parametric mapping (SPM1D) was performed within the SPM1D package based on MATLAB. IBM SPSS Statistics 26 (IBM Corporation, Armonk, NY, USA) was used to analyze sagittal ROM, sagittal PFA, peak moments, and COP for different conditions in the x- and y-axes. Two-way repeated measures were used to analyze for significance. The significance level was set at $p < 0.05$.

3. Results

3.1. Hip, Knee, and Ankle Joint Angles

The kinematic joint results and SPM1D results for the hip, knee, and ankle joints at 25% BW and 50% BW are shown in Figure 2 and Table 1.

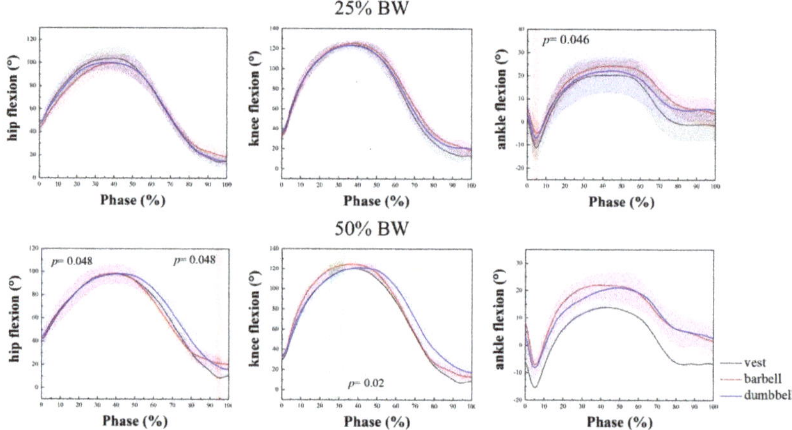

Figure 2. Joint angles and SPM1D results for hip, knee, and ankle at 25% BW and 50% BW. Red: indicates a significant difference between vest and barbell; green: indicates a significant difference between barbell and dumbbell. BW: body weight.

Table 1. Sagittal ROM and PFA of the hip, knee, and ankle at 25% BW and 50% BW.

		25% BW (Mean ± SD)		50% BW (Mean ± SD)	
		ROM	PFA	ROM	PFA
vest (°)	hip	90.04 ± 8.26	101.92 ± 9.79	88.38 ± 2.95	95.97 ± 2.57
	knee	108.98 ± 7.33	124.33 ± 3.93	110.52 ± 8.24	122.37 ± 2.37
	ankle	34.13 ± 5.66	23.51 ± 8.73 [d]	33.73 ± 5.02	20.28 ± 7.01 [d]
barbell (°)	hip	83.86 ± 12.26	101.62 ± 9.04	81.23 ± 12.68	100.23 ± 9.48
	knee	103.58 ± 6.19 [b]	125.81 ± 2.63	114.73 ± 4.21 [ab]	125.07 ± 2.80
	ankle	32.72 ± 2.98	25.50 ± 3.51 [d]	32.80 ± 4.79	23.49 ± 6.75 [d]
dumbbell (°)	hip	86.35 ± 5.47	100.23 ± 6.69	84.26 ± 10.00	99.70 ± 9.43
	knee	105.42 ± 3.56	123.31 ± 6.96	107.90 ± 3.32 [a]	122.00 ± 4.65
	ankle	29.96 ± 7.12	21.84 ± 10.44	31.40 ± 5.78	21.92 ± 7.15

Note: [a] indicates significance $p < 0.05$ for the same weight; [b] indicates significance $p < 0.05$ for different weights for the same loading type; [d] indicates the significance of the main effect of type $p < 0.05$.

Two-way repeated measures analysis was performed to compare the effects on joint ROM and PFA at different load types and load intensities. The between-group (group) effects were 25% BW and 50% BW load intensity, and the within-group (type) effects were changes in joint angles due to vest, barbell, and dumbbell load types. There were no abnormal data as judged by box-line plots. The variance–covariance matrix of the dependent variable was equal for the interaction term group × type by Mauchly's sphericity test ($p > 0.05$). For ROM and PFA in the hip joint, the group × type interaction was not statistically significant, and no significant differences were found between the main effects of group and type. ROM and PFA for vest, barbell, and dumbbell all had a decreasing trend at 50% BW compared to 25% BW. In Figure 2, in the results of SPM1D, a significant difference between the vest and barbell at 50% BW load intensity was observed in the rise and stand phase ($p = 0.048$). At ROM of the knee joint, the interaction of group × type was statistically significant, F (2, 16) = 4.744, $p = 0.024$. The separate effect of the group on the barbell was statistically significant ($p = 0.038$), and the barbell was significant at 25% BW and 50% BW ($p = 0.038$). At 50% BW, the barbell was significantly different from the dumbbell ($p = 0.033$), as shown in Table 1, and ROM increased with increasing intensity. In Table 1, PFA decreased with increasing intensity and use of the barbell decreased PFA the least, by 0.74°, whereas SPM1D results showed a significant difference between barbell and dumbbell during squatting at 50% BW deadlift strength ($p = 0.02$). Both barbell and dumbbell use increased ankle ROM, but the vest and the dumbbell increased ankle ROM more, by 1.44°. SPM1D found a significant difference between the vest and barbell at 25% BW deadlift strength when the foot touched the ground ($p = 0.046$). A significant difference in the main effect of weight types was found at PFA ($p = 0.038$). A significant difference was found between the vest and barbell ($p = 0.022$). In addition, 50% BW with the vest decreased by 3.23°, compared to 25% BW and 50% BW with the barbell which decreased by 2.01° and compared to 25% BW with dumbbells which increased by 0.08°. Overall, the barbell varied to a greater degree at the knee joint than the other two by intensity. Dumbbells varied less at the three joints than at the other two.

3.2. Hip, Knee, and Ankle Joint Moments

The joint moments of the hip, knee, and ankle joints at 25% and 50% of body weight, PFA, and SPM1D results are shown in Figure 3 and Table 2.

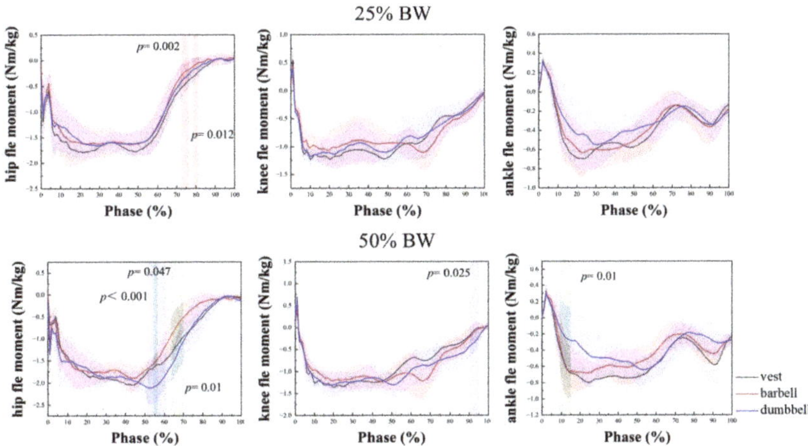

Figure 3. Joint moments and SPM1D results for hip, knee, and ankle at 25% BW and 50% BW. Red: indicates a significant difference between vest and barbell; green: indicates a significant difference between barbell and dumbbell; blue: indicates a significant difference between vest and dumbbell. BW: body weight.

Table 2. Peak joint moments of hip, knee, and ankle at 25% BW and 50% BW.

		25% BW (Mean ± SD)	50% BW (Mean ± SD)
vest (Nm/kg)	hip	−1.96 ± 0.14 [c]	−2.12 ± 0.24 [c]
	knee	−1.41 ± 0.11 [c]	−1.44 ± 0.17 [c]
	ankle	−0.79 ± 0.15 [b]	−0.98 ± 0.19 [ab]
barbell (Nm/kg)	hip	−1.87 ± 0.32 [c]	−2.09 ± 0.32 [c]
	knee	−1.46 ± 0.13 [c]	−1.59 ± 0.17 [c]
	ankle	−0.78 ± 0.18	−0.89 ± 0.22
dumbbell (Nm/kg)	hip	−1.90 ± 0.29 [c]	−2.30 ± 0.30 [c]
	knee	−1.47 ± 0.20 [c]	−1.58 ± 0.12 [c]
	ankle	−0.70 ± 0.21	−0.81 ± 0.20 [a]

Note: [a] indicates significance $p < 0.05$ for the same weight; [b] indicates significance $p < 0.05$ for different weights for the same load type; [c] indicates significance $p < 0.05$ for the main effect of the group.

As shown in Table 2, there was no significant difference in the hip group × type interaction, but there was a significant difference in the main effect of weight-bearing strength, F $(1, 11) = 93.13$, $p < 0.001$. Furthermore, 25% BW was 0.259 Nm/kg smaller than 50% BW at weight-bearing strength, a statistically significant difference ($p < 0.001$). At 25% BW, the peak barbell flexion moment was the smallest at 1.87 Nm/kg, followed by the dumbbell and the largest was the vest at 1.96 Nm/kg. Moreover, 50% BW increased the most over 25% BW with dumbbells at 0.4 Nm/kg and the least with the vest at 0.16 Nm/kg. In the SPM1D results in Figure 3, there was a significant difference between the vest and barbell rise and stand phase at 25% BW weight-bearing strength ($p = 0.002$, $p = 0.012$). At 50% BW weight-bearing strength, there was a significant difference between the barbell and dumbbell ($p = 0.047$, $p = 0.01$), and also between the vest and dumbbell ($p < 0.001$). On the knee joint, the main effect of weight-bearing strength was statistically significant, F $(1, 11) = 9.058$, $p = 0.012$. The largest increase at 50% BW over 25% BW was with the barbell, which increased by 0.13 Nm/kg from 1.46 to 1.59 Nm/kg, and the smallest was with the vest, which increased by 0.03 Nm/kg from 1.41 to 1.44 Nm/kg. In the results of SPM1D, there was a significant difference between the vest and the dumbbell in the standing phase at 50% BW ($p = 0.025$). At the ankle joint, the group × type interaction was significantly different for the ankle ($p = 0.006$), and analysis of the separate effects of group and type revealed a significant difference between the vest at 25% BW and 50% BW ($p = 0.006$). The vest increased most, from 0.79 to 0.98 Nm/kg, an increase of 0.19 Nm/kg; whereas the barbell and dumbbell both increased by 0.11 Nm/kg. At 50% BW, the vest was 0.98 Nm/kg, and the dumbbell was 0.81 Nm/kg; there was a significant difference between them ($p = 0.023$). In the SPM1D plot, it was shown that at 50% BW, there was a significant difference between the dumbbell and the barbell during the squat ($p = 0.01$). Overall, the barbell varied to a greater degree of variation in weight strength at the knee joint than the other two, and the dumbbell had relatively little overall variation at all three joints. The vest, on the other hand, showed the least variation in ROM at the hip joint.

3.3. The COP of the Lunge

The COP motion trajectory and x-axis offset range and y-axis offset at 25% BW and 50% BW are shown in Figure 4 and Table 3.

Table 3. COP x-axis offset range and y-axis offset at 25% BW and 50% BW.

	25% BW (Mean ± SD)		50% BW (Mean ± SD)	
	x-axis (mm/m)	y-axis (mm.s/m)	x-axis (mm/m)	y-axis (mm.s/m)
vest	31.07 ± 9.00	5.35 ± 1.76 [c]	35.85 ± 9.13	8.43 ± 1.94 [c]
barbell	38.08 ± 10.43 [d]	6.99 ± 3.37 [c]	35.37 ± 5.47 [d]	4.84 ± 1.68 [c]
dumbbell	28.26 ± 4.59 [d]	4.88 ± 2.54 [c]	31.75 ± 4.82 [d]	6.53 ± 2.81 [c]

Note: [c] indicates significance $p < 0.05$ for the main effect of the group. [d] indicates the significance of the main effect of type $p < 0.05$.

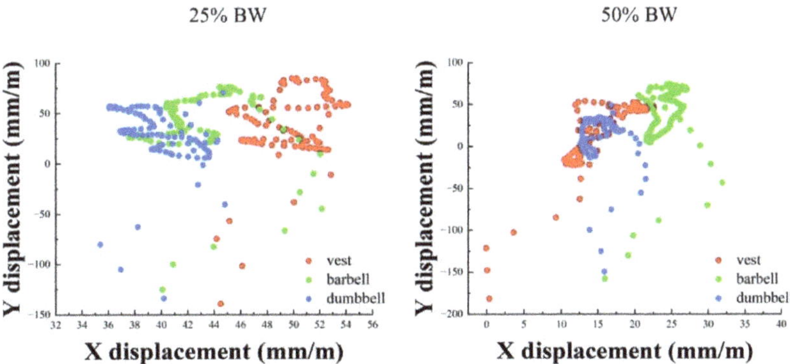

Figure 4. COP motion trajectory at 25% BW and 50% BW. BW: body weight.

As shown in Figure 4, it can be seen that the overall offset range with the dumbbell was smaller at 25% BW and 50% BW, whereas the overall trajectory with the barbell was larger at 25% BW. As shown in Table 3, in the offset range of the COP x-axis, there was no significant difference in the group × type interaction, but there was a significant difference in the main effect of weight types, F (2, 16) = 3.689, p = 0.048, and there was a statistically significant difference between barbell and dumbbell (p = 0.027). The offset range of both vest and dumbbell increased and the vest increased the most with 4.78 mm/m. In the y-axis offset, there was no significant difference in the group × type interaction. There was a significant difference in the main effect of weight strength, F (1, 9) = 25.077, p = 0.001. Additionally, 25% BW had a statistically significant difference of 2.453 mm.s/m smaller offset than 50% BW at weight strength (p = 0.001). The y-axis was the same as the x-axis, and both had the largest increase in vest offset of 3.08 mm.s/m.

4. Discussion

The purpose of this study was to investigate changes in lower limb kinematics and kinetics of male fitness novices using dumbbells, barbells, and weighted vests at 25% BW and 50% BW weights and to provide appropriate exercise choices and fitness instructions accordingly. This study found that, as strength increased, the kinematics of the barbell at the knee joint was more influential than the other two by studying different weight types and intensities, whereas the SPM1D was also used to analyze the process of joint angles and moments in this study compared to previous studies. All three weight types increased the joint moment with the increase in weight strength, and the increase in all three types was significant at the hip and knee joints, with the dumbbell increasing the most at the hip, the barbell increasing the most at the knee, and only the vest increasing significantly at the ankle joint. For COP, dumbbells showed the least variation whereas barbells decreased the COP. Weight type and intensity had a greater effect on kinetics than on kinematics, and SPM1D showed more significant results at 50% BW, which confirms our first and second hypotheses.

During the lunge phase, our study found that the knee joint had the greatest ROM and the most PFA. The ROM of the knee increased with increasing load intensity, with an average increase of 5.06° for all three load types, which is consistent with the findings of Bryan et al. [9]. The largest increase in ROM of the knee was 11.15° during barbell load conditions. In addition, a study by Danielle et al. [40] also showed an effect between greater knee flexion angle and knee joint stress during the anterior lunge. According to Zellmer et al. [41], it was reported that greater knee flexion could also lead to a tendency for the knee to translate forward. Peak kneel joint stress, quadriceps muscle strength, knee moment, knee flexion, and ankle dorsiflexion were greater when the knee was panned in front of the toes during the lunge. Keeping the knee behind the toes minimizes the stress on the patellar tendon. To reduce patellar forces, Escamilla et al. [27] suggested to be cautious about performing a long-step lunge squat. In addition to this, the anterior lunge has been shown to be part of an effective exercise program for the treatment of knee joint pain [42]. Based on the data, it is easy to see that dumbbells have a small and variable effect on the ROM of the knee relative to vests and barbells.

During the squat, we also observed a difference in joint ROM between the load cases, with SPM1D results indicating a significant difference between barbell and dumbbell at 50% BW for the knee joint. This difference may be related to the height of the entire center of gravity position. Compared to the dumbbell, the barbell's shoulder-centered high center of gravity can be difficult for novices with less physical stability, and the higher the center of gravity, the more unstable it is. Several studies have demonstrated that load weight has little effect on joint mobility [14,24]. This study found some variation at the knee joint, and this finding of the load type effect on joint ROM was through a mixture of SPM1D and load type and intensity. Changes in the process have been reported infrequently in prior studies employing this statistical methodology. At the ankle joint, we found a significant change between the vest and the barbell at 25% BW, as seen in contact with the ground and progression from plantarflexion to dorsiflexion. There was also a significant difference between the vest and barbell on the ankle PFA. According to Guilherme et al. [43], in their study, the dynamic ankle dorsiflexion range may be changed by neuromuscular strategies induced by factors outside of the passive ankle dorsiflexion range. For example, factors of hip and knee muscle weakness may be associated with an increased demand for plantar flexors. When the hip and knee muscles are weak, the lower leg and foot muscles compensate by carrying the additional load to be able to stand. Likewise, if the knee flexion is small, this indicates the presence of other underlying factors such as quadriceps weakness and limited ankle dorsiflexion range [44]. In general, the ROM with dumbbells is smaller and less variable than the other two load types, whereas overall stability is better for beginners. In contrast, those who already have the relevant basic accumulation of exercise and want to make the exercise more effective can choose the barbell method of weight bearing because the ROM and PFA at the knee joint are large and may be more sensitive to the weight.

In terms of moments, we observed that the hip joint moment was the greatest, followed by the knee joint and, finally, the ankle joint. The results of SPM1D showed that there was a significant difference between the weight types at the hip joint after reaching the lowest point and then rising at both weights. We think that this is related to the position of the center of gravity. At high load intensities, there was a significant difference between the dumbbell and the other two load types because the dumbbell load was lower than the hip position at the sides of the body. The lower the body's center of gravity, the greater the joint moment, as confirmed by the data at 50% BW. The average increase in the joint moment at 50% BW over 25% BW was 0.259 Nm/kg at the hip, 0.087 Nm/kg at the knee, and 0.139 Nm/kg at the ankle. The hip showed the greatest rise, followed by the ankle and, finally, the knee. These results confirm that the forward lunge is a hip-dominant movement and that increasing external loading has the greatest effect on the hip moment and the least on the knee joint. This varying degree of increase is consistent with the findings of Bryan [9], who argued that the knee joint has a greater range of motion than both the hip and the knee, and the examination of linear and quadratic trends in the net joint moment of the ankle suggests that the ankle may have reached the point of the maximum contribution. However, he also indicated that stronger external loading would be required to substantiate the findings. Farrokhi et al. [22] did not statistically compare hip, ankle, and knee joint dynamics. Their descriptive statistics and graphical data also showed that the forward lunge is a hip extensor-dominated movement. All of these studies indicate that the hip contributes more to the kinematics of the anterior lunge than the knee and ankle.

Bryan explained that the hip, knee, and ankle joints contribute to changes in the lunge accordingly. As the intensity of the load increases, little changes in the upper extremities or trunk, such as subjects changing the way they hold the dumbbell and moving it to a more forward position, may occur during the experiment. This phenomenon produces an effect comparable to the trunk forward lunge. The trunk forward lunge shifts the total body weight forward, and the corresponding forward movement of the barbell, dumbbell, and vest causes more mass to move forward, which may increase the contribution of the hips and ankles. Other studies have similarly pointed out that [31] as the anterior limb tibia-ground angle decreases, the ROM of the anterior knee becomes larger, and the flexion moment at the hip increases. Paul [23] suggested that due to the greater hip joint moment in the front lunge, it is more effective to choose the forward lunge over the single-leg deep squat and the back lunge if one wants to exercise the hip extensors. However, the forward shift of body weight may not be enough to ensure a meaningful increase in the hip extensors, allowing more muscle recruitment to occur in the other muscles, as mentioned by Farrokhi [22]. As seen by our data, at the hip joint, the joint moment when using the barbell at 25% BW was the smallest, followed by the dumbbell. However, the dumbbell varied greatly with increasing weight. At the knee joint, the vest produced the smallest joint moment, and at the ankle joint the dumbbell had the smallest. If we choose to exercise the hip and knee muscles more, we can choose large-weight dumbbells. If we want less load on the knee joint, we can choose a small-weight vest as the external load carrier.

For the COP displacements, it is difficult to obtain a specific understanding of the data from the point plots. The overall movement of the dumbbells was hard to distinguish, therefore we quantified the x-axis and y-axis data. The x-axis represents the displacement of the lateral COP, and the y-axis represents the displacement of the anterior and posterior aspects. We processed the y-axis data to obtain the overall offset of the y-axis. In the x-axis direction, both the vest and dumbbell increased in range with increasing weight, and there was a significant difference between the barbell and dumbbell, with an average increase of 4.78 mm/m with the vest and 3.49 mm/m with the dumbbell, whereas the y-axis offset was affected by the weight strength, with a larger offset of 2.453 mm.s/m for 50% BW than for 25% BW weight strength. The vest increased by 3.08 mm.s/m, and the dumbbell increased by 1.65 mm.s/m. This result was expected because the overall muscle strength was weak under heavy weights, and the body could not be easily controlled to complete the prescribed movements. The barbell, however, showed a decrease, as did the y-axis. We cannot explain this unexpected result specifically, probably because the upper limbs share some of the weight. However, for a wide range of displacements, Hsue et al. [35] suggested that excessive inclination leads to excessive COM and COP separation, requiring greater active postural control and more energy to counteract the increased joint moment, which would lead to loss of balance and thus accidents. Combining the above findings, it is easy to corroborate the third hypothesis that each of the three weight types has its benefits, thus selecting the correct type of weight according to the specific physical qualities of the exerciser is important.

Our study has several limitations. First, we only studied male populations; based on the relatively weak strength of female novices, we did not examine the performance of female populations in the lunge. We did not reflect the gender difference. Secondly, we only selected 25% BW and 50% BW and did not explore the results of adding weights in between or exploring the results of weights above 50% BW. Future studies should focus more on the diversity of the population, variation of the backward lunge under different weight conditions, and the effect of learning effects on results when performing different weight orders. Future studies will be more careful.

5. Conclusions

We examined different types of weight-bearing exercises. The strength of the load caused the kinematics and kinetics of the barbell to vary more at the knee joint. The kinematics of the dumbbell were less variable at all three joints, whereas the kinetics varied the most at the hip joint. The vest had the greatest effect on the joint moment at the ankle joint, whereas the joint moment changed the least at the hip and knee joints and the kinematics changed the most. All three weight-bearing types had larger joint moments with increasing intensity. These results are consistent with our expected hypothesis. One of the important findings was that all three types of weights have their own advantages for the fitness enthusiast, and training specificity and desired outcomes are important in the selection of weight-bearing exercise. Dumbbells seem to be more suitable for male beginners using the lunge exercise because of the low impact on the knee joint and high stability. For those who want to strengthen the hip and knee muscles, they can choose dumbbells with a high weight. For individuals that have accumulated experience, they can choose barbells as the main lunge exercise equipment. For people suffering from corresponding knee and knee joint pain, a small-weighted vest lunge exercise may be the best choice for rehabilitation and injury avoidance.

Author Contributions: All the authors contributed substantially to the manuscript. L.G., Z.L., M.L. and Y.G. were responsible for the conceptualization. L.G., Z.L. and J.S.B. were responsible for investigation and methodology. L.G., Z.L., M.L., J.S.B. and Y.G. were responsible for formal analysis and writing—original draft. Z.L., J.S.B. and Y.G. were responsible for writing—review and editing and supervision. All authors have read and agreed to the published version of the manuscript.

Funding: This study was supported by the Key Project of the National Social Science Foundation of China (19ZDA352), NSFC-RSE Joint Project (81911530253), Key R&D Program of Zhejiang Province China (2021C03130), Public Welfare Science and Technology Project of Ningbo, China (2021S133), Zhejiang Province Science Fund for Distinguished Young Scholars (R22A021199) and K. C. Wong Magna Fund from Ningbo University.

Institutional Review Board Statement: The study was conducted in accordance with the guidelines of the Declaration of Helsinki and was approved by the committee of Ningbo University (code RAGH202110283004.2).

Informed Consent Statement: Informed consent was obtained from all the participants involved in the study.

Data Availability Statement: The data that support the findings of this study are available upon reasonable request from the corresponding author. The data were not publicly available because of privacy or ethical restrictions.

Conflicts of Interest: The authors declare no conflict of interest.

References

1. Eliassen, W.; Saeterbakken, A.H.; van den Tillaar, R. Comparison of bilateral and unilateral squat exercises on barbell kinematics and muscle activation. *Int. J. Sports Phys. Ther.* **2018**, *13*, 871–881. [CrossRef] [PubMed]
2. Cooper, D.; Kavanagh, R.; Bolton, J.; Myers, C.; O'Connor, S. 'Prime Time of Life', A 12-Week Home-Based Online Multimodal Exercise Training and Health Education Programme for Middle-Aged and Older Adults in Laois. *Phys. Act. Health* **2021**, *5*, 178–194. [CrossRef]
3. Marchetti, P.H.; Guiselini, M.A.; da Silva, J.J.; Tucker, R.; Behm, D.G.; Brown, L.E. Balance and lower limb muscle activation between in-line and traditional lunge exercises. *J. Hum. Kinet.* **2018**, *62*, 15–22. [CrossRef] [PubMed]
4. Wu, H.W.; Tsai, C.F.; Liang, K.H.; Chang, Y.W. Effect of Loading Devices on Muscle Activation in Squat and Lunge. *J. Sport Rehabil.* **2020**, *29*, 200–205. [CrossRef]
5. Gu, Y.; Lu, Y.; Mei, Q.; Li, J.; Ren, J. Effects of different unstable sole construction on kinematics and muscle activity of lower limb. *Hum. Mov Sci.* **2014**, *36*, 46–57. [CrossRef]
6. Xu, D.; Quan, W.; Zhou, H.; Sun, D.; Baker, J.S.; Gu, Y. Explaining the differences of gait patterns between high and low-mileage runners with machine learning. *Sci. Rep.* **2022**, *12*, 2981. [CrossRef]
7. Nadzalan, A.; Mohamad, N.; Lee, J.; Chinnasee, C. Lower body muscle activation during low load versus high load forward lunge among untrained men. *J. Fundam. Appl. Sci.* **2018**, *10*, 205–217. [CrossRef]
8. Park, S.; Huang, T.; Song, J.; Lee, M. Comparative Study of the Biomechanical Factors in Range of Motion, Muscle Activity, and Vertical Ground Reaction Force between a Forward Lunge and Backward Lunge. *Phys. Ther. Rehabil. Sci.* **2021**, *10*, 98–105. [CrossRef]
9. Riemann, B.L.; Lapinski, S.; Smith, L.; Davies, G. Biomechanical analysis of the anterior lunge during 4 external-load conditions. *J. Athl. Train.* **2012**, *47*, 372–378. [CrossRef]
10. Bouillon, L.E.; Wilhelm, J.; Eisel, P.; Wiesner, J.; Rachow, M.; Hatteberg, L. Electromyographic assessment of muscle activity between genders during unilateral weight-bearing tasks using adjusted distances. *Int. J. Sports Phys. Ther.* **2012**, *7*, 595–605.
11. Ireland, M.L.; Willson, J.D.; Ballantyne, B.T.; Davis, I.M. Hip strength in females with and without patellofemoral pain. *J. Orthop. Sports Phys. Ther.* **2003**, *33*, 671–676. [CrossRef]

12. Leetun, D.T.; Ireland, M.L.; Willson, J.D.; Ballantyne, B.T.; Davis, I.M. Core stability measures as risk factors for lower extremity injury in athletes. *Med. Sci. Sports Exerc.* **2004**, *36*, 926–934. [CrossRef]

13. Powers, C.M. The influence of altered lower-extremity kinematics on patellofemoral joint dysfunction: A theoretical perspective. *J. Orthop. Sports Phys. Ther.* **2003**, *33*, 639–646. [CrossRef]

14. Lehecka, B.; Edwards, M.; Haverkamp, R.; Martin, L.; Porter, K.; Thach, K.; Sack, R.J.; Hakansson, N.A. Building a better gluteal bridge: Electromyographic analysis of hip muscle activity during modified single-leg bridges. *Int. J. Sports Phys. Ther.* **2017**, *12*, 543–549.

15. Kim, E.-K. The effect of gluteus medius strengthening on the knee joint function score and pain in meniscal surgery patients. *J. Phys. Ther. Sci.* **2016**, *28*, 2751–2753. [CrossRef]

16. Jeong, U.-C.; Sim, J.-H.; Kim, C.-Y.; Hwang-Bo, G.; Nam, C.-W. The effects of gluteus muscle strengthening exercise and lumbar stabilization exercise on lumbar muscle strength and balance in chronic low back pain patients. *J. Phys. Ther. Sci.* **2015**, *27*, 3813–3816. [CrossRef]

17. Nelson-Wong, E.; Flynn, T.; Callaghan, J.P. Development of active hip abduction as a screening test for identifying occupational low back pain. *J. Orthop. Sports Phys. Ther.* **2009**, *39*, 649–657. [CrossRef]

18. Doma, K.; Leicht, A.S.; Boullosa, D.; Woods, C.T. Lunge exercises with blood-flow restriction induces post-activation potentiation and improves vertical jump performance. *Eur. J. Appl. Physiol.* **2020**, *120*, 687–695. [CrossRef]

19. Stastny, P.; Lehnert, M.; Zaatar, A.M.; Svoboda, Z.; Xaverova, Z. Does the dumbbell-carrying position change the muscle activity in split squats and walking lunges? *J. Strength Cond. Res.* **2015**, *29*, 3177. [CrossRef]

20. Nayanti, A.P.; Prabowo, T.; Sari, D.M. The Effects of Kinesio Taping and Quadriceps Muscle Strengthening Exercise on Quadriceps Muscle Strength and Functional Status in Knee Osteoarthritis. *J. Med. Health* **2020**, *2*, 40–50. [CrossRef]

21. Bobbert, M.F.; Van Zandwijk, J. Dynamics of force and muscle stimulation in human vertical jumping. *Med. Sci. Sports Exerc.* **1999**, *31*, 303–310. [CrossRef]

22. Farrokhi, S.; Pollard, C.D.; Souza, R.B.; Chen, Y.J.; Reischl, S.; Powers, C.M. Trunk position influences the kinematics, kinetics, and muscle activity of the lead lower extremity during the forward lunge exercise. *J. Orthop. Sports Phys. Ther.* **2008**, *38*, 403–409. [CrossRef]

23. Comfort, P.; Jones, P.A.; Smith, L.C.; Herrington, L. Joint kinetics and kinematics during common lower limb rehabilitation exercises. *J. Athl. Train.* **2015**, *50*, 1011–1018. [CrossRef]

24. Schütz, P.; List, R.; Zemp, R.; Schellenberg, F.; Taylor, W.R.; Lorenzetti, S. Joint angles of the ankle, knee, and hip and loading conditions during split squats. *J. Appl. Biomech.* **2014**, *30*, 373–380. [CrossRef]

25. Phomsoupha, M.; Laffaye, G. The science of badminton: Game characteristics, anthropometry, physiology, visual fitness and biomechanics. *Sports Med.* **2015**, *45*, 473–495. [CrossRef]

26. Cronin, J.; McNair, P.J.; Marshall, R.N. Lunge performance and its determinants. *J. Sports Sci.* **2003**, *21*, 49–57. [CrossRef]

27. Escamilla, R.F.; Zheng, N.; MacLeod, T.D.; Edwards, W.B.; Hreljac, A.; Fleisig, G.S.; Wilk, K.E.; Moorman III, C.T.; Imamura, R.; Andrews, J.R. Patellofemoral joint force and stress between a short-and long-step forward lunge. *J. Orthop. Sports Phys. Ther.* **2008**, *38*, 681–690. [CrossRef]

28. Muyor, J.M.; Martin-Fuentes, I.; Rodriguez-Ridao, D.; Antequera-Vique, J.A. Electromyographic activity in the gluteus medius, gluteus maximus, biceps femoris, vastus lateralis, vastus medialis and rectus femoris during the Monopodal Squat, Forward Lunge and Lateral Step-Up exercises. *PLoS ONE* **2020**, *15*, e0230841. [CrossRef] [PubMed]

29. Dill, K.E.; Begalle, R.L.; Frank, B.S.; Zinder, S.M.; Padua, D.A. Altered knee and ankle kinematics during squatting in those with limited weight-bearing-lunge ankle-dorsiflexion range of motion. *J. Athl. Train.* **2014**, *49*, 723–732. [CrossRef] [PubMed]

30. Escamilla, R.F.; Zheng, N.; MacLeod, T.D.; Imamura, R.; Edwards, W.B.; Hreljac, A.; Fleisig, G.S.; Wilk, K.E.; Moorman III, C.T.; Paulos, L. Cruciate ligament tensile forces during the forward and side lunge. *Clin. Biomech.* **2010**, *25*, 213–221. [CrossRef] [PubMed]

31. Ying, S.; Li, H. Reliability and Validity of the Short Version of Perceived Benefits and Barriers Scale for Physical Activity. *Phys. Act. Health* **2022**, *6*, 64–72. [CrossRef]

32. Macrum, E.; Bell, D.R.; Boling, M.; Lewek, M.; Padua, D. Effect of limiting ankle-dorsiflexion range of motion on lower extremity kinematics and muscle-activation patterns during a squat. *J. Sport Rehabil.* **2012**, *21*, 144–150. [CrossRef]

33. Shi, Z.; Sun, D. Conflict between Weightlifting and Health? The Importance of Injury Prevention and Technology Assistance. *Phys. Act. Health* **2022**, *6*, 1–4. [CrossRef]

34. Xu, D.; Jiang, X.; Cen, X.; Baker, J.S.; Gu, Y. Single-leg landings following a volleyball spike may increase the risk of anterior cruciate ligament injury more than landing on both-legs. *Appl. Sci.* **2020**, *11*, 130. [CrossRef]

35. Hsue, B.-J.; Miller, F.; Su, F.-C. The dynamic balance of the children with cerebral palsy and typical developing during gait. Part I: Spatial relationship between COM and COP trajectories. *Gait Posture* **2009**, *29*, 465–470. [CrossRef]

36. Zhou, H.; Xu, D.; Quan, W.; Liang, M.; Ugbolue, U.C.; Baker, J.S.; Gu, Y. A Pilot Study of Muscle Force between Normal Shoes and Bionic Shoes during Men Walking and Running Stance Phase Using Opensim. *Actuators* **2021**, *10*, 274. [CrossRef]

37. Lu, Y.; Mei, Q.; Peng, H.T.; Li, J.; Wei, C.; Gu, Y. A Comparative Study on Loadings of the Lower Extremity during Deep Squat in Asian and Caucasian Individuals via OpenSim Musculoskeletal Modelling. *BioMed Res. Int.* **2020**, *2020*, 7531719. [CrossRef]

38. Catelli, D.S.; Wesseling, M.; Jonkers, I.; Lamontagne, M. A musculoskeletal model customized for squatting task. *Comput. Methods Biomech. Biomed. Eng.* **2019**, *22*, 21–24. [CrossRef]

39. Butler, A.B.; Caruntu, D.I.; Freeman, R.A. Knee joint biomechanics for various ambulatory exercises using inverse dynamics in OpenSim. In Proceedings of the ASME International Mechanical Engineering Congress and Exposition, Tampa, FL, USA, 3–9 November 2017; p. V003T004A045.
40. Goulette, D.; Griffith, P.; Schiller, M.; Rutherford, D.; Kernozek, T.W. Patellofemoral joint loading during the forward and backward lunge. *Phys. Ther. Sport* **2021**, *47*, 178–184. [CrossRef]
41. Zellmer, M.; Kernozek, T.W.; Gheidi, N.; Hove, J.; Torry, M. Patellar tendon stress between two variations of the forward step lunge. *J. Sport Health Sci.* **2019**, *8*, 235–241. [CrossRef]
42. Xiang, L.; Mei, Q.; Wang, A.; Shim, V.; Fernandez, J.; Gu, Y. Evaluating function in the hallux valgus foot following a 12-week minimalist footwear intervention: A pilot computational analysis. *J. Biomech.* **2022**, *132*, 110941. [CrossRef]
43. da Costa, G.V.; de Castro, M.P.; Sanchotene, C.G.; Ribeiro, D.C.; de Brito Fontana, H.; Ruschel, C. Relationship between passive ankle dorsiflexion range, dynamic ankle dorsiflexion range and lower limb and trunk kinematics during the single-leg squat. *Gait Posture* **2021**, *86*, 106–111. [CrossRef]
44. Horan, S.A.; Watson, S.L.; Carty, C.P.; Sartori, M.; Weeks, B.K. Lower-limb kinematics of single-leg squat performance in young adults. *Physiother. Can.* **2014**, *66*, 228–233. [CrossRef]

 *bioengineering*

Article

Dynamics of Two-Link Musculoskeletal Chains during Fast Movements: Endpoint Force, Axial, and Shear Joint Reaction Forces

Andrea Biscarini 

Department of Medicine and Surgery, University of Perugia, 06132 Perugia, Italy; andrea.biscarini@unipg.it;
Tel.: +39-075-5858135

Abstract: This study provides a dynamic model for a two-link musculoskeletal chain controlled by single-joint and two-joint muscles. The chain endpoint force, and the axial and shear components of the joint reaction forces, were expressed analytically as a function of the muscle forces or torques, the chain configuration, and the link angular velocities and accelerations. The model was applied to upper-limb ballistic push movements involving transverse plane shoulder flexion and elbow extension. The numerical simulation highlights that the shoulder flexion and elbow extension angular acceleration at the initial phase of the movement, and the elbow extension angular velocity and acceleration at the later phase of the movement, induce a proportional medial deviation in the endpoint force direction. The forearm angular velocity and acceleration selectively affect the value of the axial and shear components of the shoulder reaction force, depending on the chain configuration. The same goes for the upper arm and elbow. The combined contribution of the elbow extension angular velocity and acceleration may give rise to anterior shear force acting on the humerus and axial forearm traction force as high as 300 N. This information can help optimize the performance and estimate/control of the joint loads in ballistic sport activities and power-oriented resistance exercises.

Keywords: biomechanics; kinetic chain; joint loads; fast movement; ballistic exercise

Citation: Biscarini, A. Dynamics of Two-Link Musculoskeletal Chains during Fast Movements: Endpoint Force, Axial, and Shear Joint Reaction Forces. *Bioengineering* **2023**, *10*, 240. https://doi.org/10.3390/bioengineering10020240

Academic Editors: Rui Zhang, Wei-Hsun Tai and Aurélien Courvoisier

Received: 24 November 2022
Revised: 4 February 2023
Accepted: 8 February 2023
Published: 11 February 2023

1. Introduction

The planar system, composed of a fixed frame and two rigid bodies connected serially by ideal revolute joints controlled by torque actuators, is a reference model in robotics [1,2] and musculoskeletal biomechanics [3–6]. For the sake of brevity, throughout this paper, the term "chain" (or "serial chain") is used to refer to this system, and the term "link" is used to refer to each rigid body component in the chain [1–4]. The mechanics of this ideal model have been characterized by the direct and inverse kinematic equations (establishing the relationship between the kinematic variables of the joints and those of the free endpoint of the chain), the corresponding direct and inverse static equations (determining the equilibrating joint torques necessary to maintain the chain equilibrium when an external force acts on the endpoint of the chain, and, inversely, the force exerted by the endpoint on a fixed external environment given the joint torques), the apparent endpoint and joint stiffness, and the closed-form dynamic equations (the dynamic equations that contain all the variables in explicit input–output form) [1–4].

The static equations for the two-link chain have been applied to the study of human limb biomechanics, considering the chain composed of the shoulder and elbow joints, the upper arm and the forearm-hand links, and the trunk as a fixed base or the corresponding lower limb chain [4]. Notably, previous static analyses established that the torque acting on the proximal joint (shoulder and hip) produces endpoint force along the pointing axis, which is the longitudinal axis of the distal link (forearm and lower leg), whereas the torque acting on the distal joint (elbow and knee) produces endpoint force along the radial axis, which is the axis connecting the proximal joint (shoulder and hip) with the chain endpoint (hand and foot) [4,7]. This information can potentially help therapists and athletic trainers analyze and optimize the technique of strength-training exercises, and estimate, even possibly control, the reaction forces acting on specific joint structures during these exercises [4].

Critically, the direct and inverse static equations can only be applied to isometric or quasi-static exercises, thus, limiting their potential application in exercise and sport sciences. During explosive motor actions (e.g., baseball pitch, team handball throw, tennis serve, volleyball spike, or shot put), the elbow extension reaches peak angular velocities between 1300 and 2300 $^\circ$/s, and peak angular accelerations as high as $10^4 \div 4 \cdot 10^4$ $^\circ$/s^2 [8–12]. High elbow angular velocity (~850 $^\circ$/s) can also be recorded during power-oriented exercises, such as the explosive bench press throws executed with low to moderate resistance (30–50% of one repetition maximum) [13]. In these fast dynamic conditions, the quasi-static computational approach is not applicable, and the dynamic equations of the chain should be used to determine the joint reaction forces and the relationship between the joint torques and the endpoint force.

The joints of a real musculoskeletal chain are driven by muscles that can span multiple joints, rather than ideal torque actuators. An ideal torque actuator solely produces torque about the joint axis with no effect on the joint reaction force [4]. In contrast, a muscle can produce torques among all the joints it crosses and can induce axial and shear joint reaction forces on each of these joints [14]. The selective determinations of the axial and shear reaction forces acting on a skeletal joint are fundamental to estimating how these forces are distributed among the different joint structures and periarticular tissues. The determination and control of such forces are some of the most crucial issues in musculoskeletal biomechanics [15–23].

To overcome the above limitations, this study provides a dynamic model for analysis of the fast movements of a planar two-link musculoskeletal chain driven by single-joint and two-joint muscles, modeled as force actuators attached to the links and the base. The study is aimed at deriving the analytical expression of the force exerted by the endpoint on the external environment as a function of the joint torques developed by muscles, the chain configuration, and the angular velocities and accelerations of the links. Numerical simulations were carried out considering a chain structure modelling the upper limb and dynamic chain configurations of particular interest in sports and exercise science. Further, the study investigated how the link angular velocities and accelerations contribute, in addition to the endpoint and muscle forces, to the development of the axial and shear components of the reaction forces acting on the joints. We hypothesize that the elbow and shoulder angular velocities and accelerations typically reached during ballistic sport movements can yield a relevant contribution to both the axial and shear reaction forces acting on the shoulder and elbow.

2. Materials and Methods

We consider a planar musculoskeletal chain composed of 2 slender rigid links and ideal revolute joints (Figure 1). The proximal link of the chain (link 1) is articulated with the distal link (link 2) at joint J_2 and with a fixed base (link 0) at joint J_1. The free extremity of the distal link is referred to as the endpoint P of the chain. Single-joint muscles (connecting link 1 to link 2, or link 1 to the base) and biarticular muscles (connecting link 2 to the base) are modeled as linear force actuators. The following parameters are associated with each of the two links i ($i = 1, 2$) of the chain: the mass m_i, the length l_i, the distance $l_{C_i} = |J_i C_i|$ from J_i of the center of mass C_i of the link, the link unit vector $\hat{u}_i$ pointing from J_i along the link direction, the angle θ_i between $\hat{u}_i$ and the horizontal x-axis (the counterclockwise direction of rotation is considered positive), the moments of inertia I_{J_i} about J_i, the angular velocity $\vec{\omega}_i = \dot{\theta}_i \hat{k}$ and acceleration $\dot{\vec{\omega}}_i = \ddot{\theta}_i \hat{k}$ ($\hat{k}$ is the unit vector normal to the plane of the chain), and the unit vector $\hat{w}_i = \hat{k} \times \hat{u}_i$.

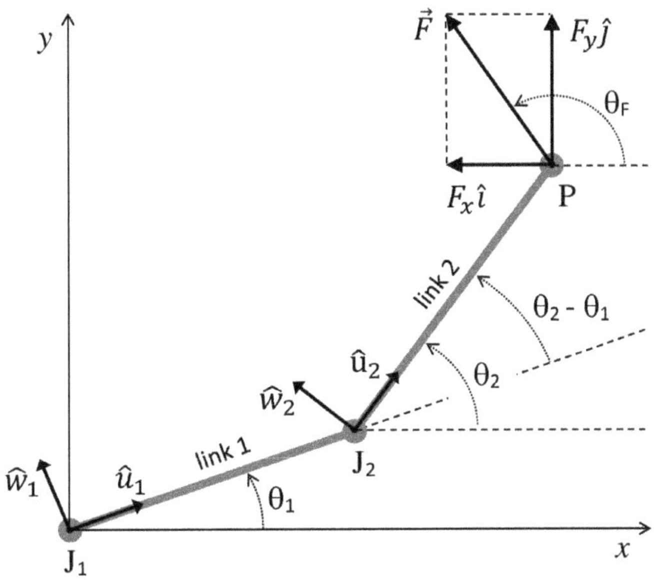

Figure 1. Mechanical diagram of the two-link chain with joint J_1 and J_2. The diagram includes the force $\vec{F} = F_x\hat{i} + F_y\hat{j}$ that the chain endpoint P exerts on the environment and the relevant angular quantities and unit vectors associated with each of the two links. The unit vectors of the x and y axis are $\hat{i}$ and $\hat{j}$, respectively.

The external forces acting on link 1 are the weight $m_1\vec{g}$ of the link applied at C_1, the muscle forces $\sum_m \vec{F}_{12}^{\,m}$ and $\sum_m \vec{F}_{10}^{\,m}$ exerted on link 1 by the muscles connecting link 1 to link 2 and to link 0, respectively, and the joint reaction forces $\vec{\phi}_{1,0}$ and $\vec{\phi}_{1,2}$ exerted on link 1 by the base (link 0) and link 2, respectively. The external forces acting on link 2 are the weight $m_2\vec{g}$ of the link applied at C_2, the muscle forces $\sum_m \vec{F}_{21}^{\,m}$ and $\sum_m \vec{F}_{20}^{\,m}$ exerted on link 2 by the muscles connecting link 2 to link 1 and to link 0, respectively, the joint reaction force $\vec{\phi}_{2,1}$ exerted on link 2 by link 1, and a contact external resistance $\vec{R}$ applied at the endpoint P of the chain. R_x and R_y denote the components of $\vec{R}$ relative to the x- and y-axis of the adopted global reference system $J_1 xy$ displayed in Figure 1. The above $\sum_m \vec{F}_{ij}^{\,m}$ summations are extended over all muscles joining link i to link j, and P_{ij}^m will denote the point of application of $\vec{F}_{ij}^{\,m}$. The following relations hold: $\vec{\phi}_{i,j} = -\vec{\phi}_{j,i}$ and $\vec{F}_{ij}^{\,m} = -\vec{F}_{ji}^{\,m}$.

The moment equations for a planar chain composed of two slender links can be derived as a particular case of the general moment equations from Biscarini (2021) [14] for non-slender n-link chains driven by single-joint and multi-joint muscle actuators:

$$\tau_2 = m_2 g\, l_{C_2} \cos\theta_2 + R_x l_2 \sin\theta_2 - R_y l_2 \cos\theta_2 + I_{J_2}\ddot{\theta}_2 \\ + m_2 l_{C_2} l_1 \left[\ddot{\theta}_1 \cos(\theta_2 - \theta_1) + \dot{\theta}_1^2 \sin(\theta_2 - \theta_1)\right] \tag{1}$$

$$\tau_1 = \tau_2 + m_1 g\, l_{C_1} \cos\theta_1 + m_2 g l_1 \cos\theta_1 + R_x l_1 \sin\theta_1 - R_y l_1 \cos\theta_1 \\ + \left(I_{J_1} + m_2 l_1^2\right)\ddot{\theta}_1 + l_1 m_2 l_{C_2}\left[\ddot{\theta}_2 \cos(\theta_2 - \theta_1) - \dot{\theta}_2^2 \sin(\theta_2 - \theta_1)\right] \tag{2}$$

here, τ_1 (τ_2) is the torque developed about J_1 (J_2) by the distal to proximal forces of all single-joint and two-joint muscles crossing J_1 (J_2):

$$\tau_1 = \left(\sum_m J_1 P_{10}^m \times \vec{F}_{10}^{\,m} + \sum_m J_1 P_{20}^m \times \vec{F}_{20}^{\,m} \right) \cdot \hat{k} \tag{3}$$

$$\tau_2 = \left(\sum_m J_2 P_{21}^m \times \vec{F}_{21}^{\,m} + \sum_m J_2 P_{20}^m \times \vec{F}_{20}^{\,m} \right) \cdot \hat{k} \tag{4}$$

Moments are considered positive if they produce angular acceleration in a counterclockwise direction ($\hat{k}$ is the unit vector of the z-axis, normal to the plane of the chain). The joint reaction forces $\vec{\phi}_{1,2} = -\vec{\phi}_{2,1}$ and $\vec{\phi}_{0,1} = -\vec{\phi}_{0,1}$ can also be derived by [14]:

$$\vec{\phi}_{1,2} = m_2 \vec{g} + R_x \hat{i} + R_y \hat{j} + \sum_m \vec{F}_{21}^{\,m} + \sum_m \vec{F}_{20}^{\,m} - m_2 \left[l_1 \left(\ddot{\theta}_1 \hat{w}_1 - \dot{\theta}_1^2 \hat{u}_1 \right) + l_{C_2} \left(\ddot{\theta}_2 \hat{w}_2 - \dot{\theta}_2^2 \hat{u}_2 \right) \right] \tag{5}$$

$$\vec{\phi}_{0,1} = m_1 \vec{g} + m_2 \vec{g} + R_x \hat{i} + R_y \hat{j} + \sum_m \vec{F}_{10}^{\,m} + \sum_m \vec{F}_{20}^{\,m} - m_2 \left[l_1 \left(\ddot{\theta}_1 \hat{w}_1 - \dot{\theta}_1^2 \hat{u}_1 \right) + l_{C_2} \left(\ddot{\theta}_2 \hat{w}_2 - \dot{\theta}_2^2 \hat{u}_2 \right) \right]$$
$$- m_1 l_{C_1} \left(\ddot{\theta}_1 \hat{w}_1 - \dot{\theta}_1^2 \hat{u}_1 \right) \tag{6}$$

In Equations (1), (2), (5), and (6), different from the general case [14], we considered an external resistance $\vec{R}$ that only acts at the endpoint P of the chain; $\vec{R}$ is expressed in terms of its Cartesian components R_x and R_y.

Equations (1) and (2) can be rewritten as a linear system in the unknowns R_x and R_y:

$$l_1 \sin \theta_1 R_x - l_1 \cos \theta_1 R_y$$
$$= \tau_1 - \tau_2 - m_1 g \, l_{C_1} \cos \theta_1 - m_2 g l_1 \cos \theta_1 - \left(I_{J_1} + m_2 l_1^2 \right) \ddot{\theta}_1$$
$$- l_1 m_2 l_{C_2} \left[\ddot{\theta}_2 \cos(\theta_2 - \theta_1) - \dot{\theta}_2^2 \sin(\theta_2 - \theta_1) \right] \tag{7}$$

$$l_2 \sin \theta_2 R_x - l_2 \cos \theta_2 R_y = \tau_2 - m_2 g \, l_{C_2} \cos \theta_2 - I_{J_2} \ddot{\theta}_2 - m_2 l_{C_2} l_1 \left[\ddot{\theta}_1 \cos(\theta_2 - \theta_1) + \dot{\theta}_1^2 \sin(\theta_2 - \theta_1) \right]$$

According to Newton's third law, the force $\vec{F}$ exerted by the endpoint P on the external environment that provides the external resistance $\vec{R}$ is given by $\vec{F} = -\vec{R}$. Thus, the $\vec{F}$ components are given by:

$$\begin{cases} F_x = -R_x \\ F_y = -R_y \end{cases} \tag{8}$$

In the case of external resistance of the inertial type, $\vec{R}$, and consequently $\vec{F}$, include the inertial force components related to the resistance mass acceleration. The solution of the Equation system (7) determines R_x and R_y, then Equation (8) gives the analytical expressions of the and F_y components of the force exerted by the chain endpoint:

$$F_x = \frac{l_2 \cos \theta_2}{l_1 l_2 \sin(\theta_2 - \theta_1)} \{ \tau_1 - \tau_2 - m_1 g \, l_{C_1} \cos \theta_1 - m_2 g l_1 \cos \theta_1 - \left(I_{J_1} + m_2 l_1^2 \right) \ddot{\theta}_1$$
$$- l_1 m_2 l_{C_2} \left[\ddot{\theta}_2 \cos(\theta_2 - \theta_1) - \dot{\theta}_2^2 \sin(\theta_2 - \theta_1) \right] \}$$
$$- \frac{l_1 \cos \theta_1}{l_1 l_2 \sin(\theta_2 - \theta_1)} \{ \tau_2 - m_2 g \, l_{C_2} \cos \theta_2 - I_{J_2} \ddot{\theta}_2$$
$$- m_2 l_{C_2} l_1 \left[\ddot{\theta}_1 \cos(\theta_2 - \theta_1) + \dot{\theta}_1^2 \sin(\theta_2 - \theta_1) \right] \} \tag{9}$$

$$F_y = \frac{l_2 \sin\theta_2}{l_1 l_2 \sin(\theta_2 - \theta_1)} \Big\{ \tau_1 - \tau_2 - m_1 g\, l_{C_1} \cos\theta_1 - m_2 g l_1 \cos\theta_1$$

$$- \left(I_{J_1} + m_2 l_1^2 \right) \ddot{\theta}_1$$

$$- l_1 m_2 l_{C_2} \left[\ddot{\theta}_2 \cos(\theta_2 - \theta_1) - \dot{\theta}_2^2 \sin(\theta_2 - \theta_1) \right] \Big\}$$

$$- \frac{l_1 \sin\theta_1}{l_1 l_2 \sin(\theta_2 - \theta_1)} \Big\{ \tau_2 - m_2 g\, l_{C_2} \cos\theta_2 - I_{J_2} \ddot{\theta}_2$$

$$- m_2 l_{C_2} l_1 \left[\ddot{\theta}_1 \cos(\theta_2 - \theta_1) + \dot{\theta}_1^2 \sin(\theta_2 - \theta_1) \right] \Big\}$$

(10)

The direction of $\vec{F}$ is determined by the following equation:

$$\theta_F = \arctan\left(\frac{F_y}{F_x} \right) \tag{11}$$

where θ_F is the angle between $\vec{F}$ and the x-axis (Figure 1).

The axial and shear components of the joint reaction force $\vec{\phi}_{1,2}$ acting on J_2 are determined by projecting $\vec{\phi}_{1,2}$ on the direction of $\hat{u}_2$ and $\hat{w}_2$, respectively:

$$\begin{aligned}
\left(\phi_{1,2} \right)_{\text{axial}} &= \vec{\phi}_{1,2} \cdot \hat{u}_2 \\
&= R_x \cos\theta_2 + \left(R_y - m_2 g \right) \sin\theta_2 + \sum_m \vec{F}_{21}^{\,m} \cdot \hat{u}_2 + \sum_m \vec{F}_{20}^{\,m} \cdot \hat{u}_2 \\
&\quad + m_2 l_{C_2} \dot{\theta}_2^2 + m_2 l_1 \dot{\theta}_1^2 \cos(\theta_2 - \theta_1) - m_2 l_1 \ddot{\theta}_1 \sin(\theta_2 - \theta_1)
\end{aligned} \tag{12}$$

$$\begin{aligned}
\left(\phi_{1,2} \right)_{\text{shear}} &= \vec{\phi}_{1,2} \cdot \hat{w}_2 \\
&= -R_x \sin\theta_2 + \left(R_y - m_2 g \right) \cos\theta_2 + \sum_m \vec{F}_{21}^{\,m} \cdot \hat{w}_2 + \sum_m \vec{F}_{20}^{\,m} \cdot \hat{w}_2 \\
&\quad - m_2 l_{C_2} \ddot{\theta}_2 - m_2 l_1 \dot{\theta}_1^2 \sin(\theta_2 - \theta_1) - m_2 l_1 \ddot{\theta}_1 \cos(\theta_2 - \theta_1)
\end{aligned} \tag{13}$$

Likewise, the axial and shear components of the joint reaction force $\vec{\phi}_{0,1}$ acting on J_1 are determined by projecting $\vec{\phi}_{0,1}$ on the direction of $\hat{u}_1$ and $\hat{w}_1$, respectively:

$$\begin{aligned}
\left(\phi_{0,1} \right)_{\text{axial}} &= \vec{\phi}_{0,1} \cdot \hat{u}_1 \\
&= R_x \cos\theta_1 + \left(R_y - m_1 g - m_2 g \right) \sin\theta_1 + \sum_m \vec{F}_{10}^{\,m} \cdot \hat{u}_1 + \sum_m \vec{F}_{20}^{\,m} \cdot \hat{u}_1 + \left(m_1 l_{C_1} + m_2 l_1 \right) \dot{\theta}_1^2 \\
&\quad + m_2 l_{C_2} \dot{\theta}_2^2 \cos(\theta_2 - \theta_1) + m_2 l_{C_2} \ddot{\theta}_2 \sin(\theta_2 - \theta_1)
\end{aligned} \tag{14}$$

$$\begin{aligned}
\left(\phi_{0,1} \right)_{\text{shear}} &= \vec{\phi}_{0,1} \cdot \hat{w}_1 \\
&= -R_x \sin\theta_1 + \left(R_y - m_1 g - m_2 g \right) \cos\theta_1 + \sum_m \vec{F}_{10}^{\,m} \cdot \hat{w}_1 + \sum_m \vec{F}_{20}^{\,m} \cdot \hat{w}_1 - \left(m_1 l_{C_1} + m_2 l_1 \right) \ddot{\theta}_1 \\
&\quad + m_2 l_{C_2} \dot{\theta}_2^2 \sin(\theta_2 - \theta_1) - m_2 l_{C_2} \ddot{\theta}_2 \cos(\theta_2 - \theta_1)
\end{aligned} \tag{15}$$

In the following numerical simulation, F_x, F_y, and θ_F (given by Equations (9)–(11)) are regarded as functions of the 8 angular and torque variables τ_1, τ_2, θ_1, θ_2, $\dot{\theta}_1, \dot{\theta}_2, \ddot{\theta}_1$, and $\ddot{\theta}_2$. To highlight the effect of these variables, the weight of the links is neglected (this corresponds to chain movements in the horizontal plane) and the geometric and inertial quantities l_1, l_2, l_{C_1}, l_{C_2}, m_1, m_2, I_{J_1}, and I_{J_2} are considered fixed parameters. Their values were retrieved from the anatomic data reported by Winter [24], considering the chain composed of the shoulder (S) and elbow (E) joints ($J_1 = S$, $J_2 = E$), the upper arm (link 1) and the forearm hand (link 2), and the stationary trunk (fixed base). With this musculoskeletal chain, τ_1 and τ_2 are the shoulder and elbow muscle torque, respectively. An upper limb push movement in the transverse plane will be specifically analyzed. By convention, positive (negative) values of τ_1 reflect a shoulder flexion (extension) muscle torque in that plane, while positive (negative) values of τ_2 reflect an elbow flexion (extension) muscle torque.

The values of F_x, F_y, and θ_F were analyzed for specific subsets of values of τ_1, τ_2, θ_1, θ_2, $\dot{\theta}_1$, $\dot{\theta}_2$, $\ddot{\theta}_1$, and $\ddot{\theta}_2$ with physiological meaning and specific interest in human movement and physical exercise:

- Condition 1. Effect of the upper arm angular acceleration $\ddot{\theta}_1$ (which coincides with the shoulder flexion angular acceleration) and the shoulder flexion muscle torque τ_1 ($\tau_1 > 0$, $\tau_2 \sim 0$) at the beginning phase of an upper limb push movement, with the forearm maintained in forward direction ($\theta_1 = -30°$, $\theta_2 = 90°$, $\dot{\theta}_1 = \dot{\theta}_2 = \ddot{\theta}_2 = 0$). In this condition, the elbow extension angular acceleration is equal in magnitude to the shoulder flexion angular acceleration.

- Condition 2. Effect of the forearm angular velocity $\dot{\theta}_2$ and acceleration $\ddot{\theta}_2$, and the elbow extension muscle toque τ_2 ($\tau_2 < 0$, $\tau_1 \sim 0$), at the subsequent phase of the movement, when $\theta_1 = 60°$, $\theta_2 = 120°$, and the upper arm angular velocity and acceleration are negligible compared to those of the forearm. In this condition, $\dot{\theta}_2$ and $\ddot{\theta}_2$ coincide approximately with the elbow angular velocity and acceleration.

In these two conditions, the contributions of the link angular velocities and accelerations to the axial and shear components of the joint reaction forces $\vec{\phi}_{0,1}$ and $\vec{\phi}_{1,2}$ were determined by Equations (12)–(15). These contributions might be relevant as values of joint angular velocity and acceleration as high as 2000 $°/s$ and 20,000 $°/s^2$, respectively, have often been recorded in high-velocity upper limb sport gestures [8–12]. Maple computing software (Maplesoft, Waterloo, ON, Canada) was used to perform numerical simulations.

3. Results

3.1. Condition 1

An increase in τ_1 yields a linear increase in the vertical component F_y of the endpoint force $\vec{F}$ without affecting the horizontal component F_x of $\vec{F}$ (Figure 2a). As expected, in the static condition ($\ddot{\theta}_1 = \dot{\theta}_1 = 0$), $F_x = 0$ and $\vec{F}$ is directed along the pointing axis ($\theta_F = 90°$). An increase in $\ddot{\theta}_1$ yields a linear decrease in both F_x (which becomes negative) and F_y (Figure 2a). As a result, the angle θ_F turns out to be a non-linear increasing function of $\ddot{\theta}_1$ and non-linear decreasing function of τ_1, always greater than 90° for $\ddot{\theta}_1 > 0$ (Figure 2b). For the selected chain configuration ($\theta_1 = -30°$, $\theta_2 = 90°$), the values of θ_F may only range from 30° to 210°. In fact, for $210° < \theta_F < 360°$ and $0° < \theta_F < 30°$, the external resistance $\vec{R}$ acting on the endpoint ($\vec{R} = -\vec{F}$) would produce a horizontal shoulder flexion torque, changing the nature of the exercise from push to pull.

The shoulder acceleration $\ddot{\theta}_1$ produces negative shear force $(\phi_{0,1})_{\text{shear}}$ and axial force $(\phi_{1,2})_{\text{axial}}$, as well as positive shear force $(\phi_{1,2})_{\text{shear}}$. These three force components increase linearly in magnitude with $\ddot{\theta}_1$ (Figure 2c). Since $\vec{\phi}_{i-1,i}$ is the force exerted by the distal link (link i) on the adjacent proximal link (link $i - 1$) through the joint J_i, this means that, due to the shoulder acceleration $\ddot{\theta}_1$, the humerus is subjected to a posterior shear force and the forearm to anterior shar force and axial compression. No axial force is induced by $\ddot{\theta}_1$ on the humerus.

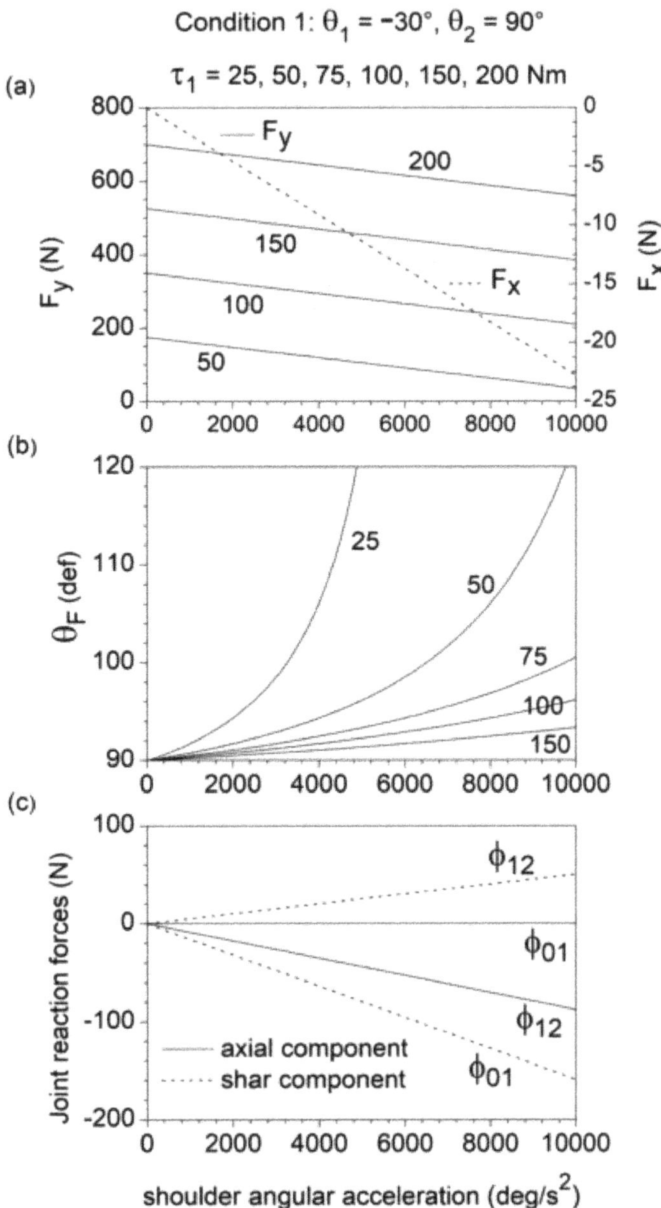

Figure 2. Dependence on the humerus angular acceleration $\ddot{\theta}_1$ of the components F_x and F_y of the endpoint force $\vec{F}$ (**a**), the direction θ_F of $\vec{F}$ (**b**), and the shear and axial components of the joint reaction forces $(\phi_{0,1})_{\text{axial}}$, $(\phi_{0,1})_{\text{shear}}$, $(\phi_{1,2})_{\text{axial}}$, and $(\phi_{1,2})_{\text{shear}}$ (**c**), for $\theta_1 = -30°$, $\theta_2 = 90°$ and $\dot{\theta}_1 = \dot{\theta}_2 = \ddot{\theta}_2 = 0$ (initial phase of upper limb push movement). In this condition, $\ddot{\theta}_1$ coincides with the shoulder angular acceleration. F_x, F_y, and θ_F are reported for different values of shoulder muscle torque τ_1 and negligible elbow muscle torque ($\tau_2 = 0$). Only the selective effect of $\ddot{\theta}_1$ on the joint reaction forces was considered in (**c**).

3.2. Condition 2

An increase in the elbow extensor torque τ_2 from -25 Nm to -75 Nm enhances F_y without affecting F_x (Figures 3a and 4a), thus, decreasing θ_F (Figures 3b and 4b). θ_F (F_x) turns out to be an increasing (decreasing) function of both $\left|\dot{\theta}_2\right|$ and $\left|\ddot{\theta}_2\right|$ (negative values of $\dot{\theta}_2$ and $\ddot{\theta}_2$ correspond to elbow extension angular velocity and acceleration, respectively). The F_y component is slightly affected by $\ddot{\theta}_2$ (Figure 3a) but displays a non-liner increasing dependence on $\left|\dot{\theta}_2\right|$ (Figure 4a).

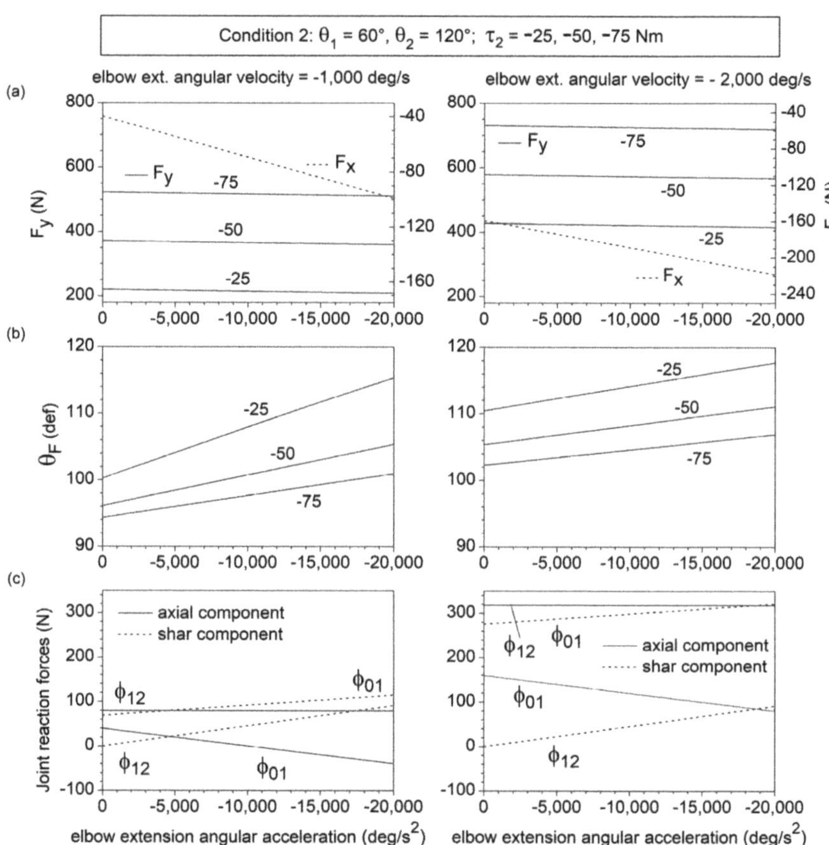

Figure 3. Dependence on the forearm angular acceleration $\ddot{\theta}_2$ of the components F_x and F_y of the endpoint force $\vec{F}$ (**a**), the direction θ_F of $\vec{F}$ (**b**), and the shear and axial components of the joint reaction forces $(\phi_{0,1})_{\text{axial}}$, $(\phi_{0,1})_{\text{shear}}$, $(\phi_{1,2})_{\text{axial}}$, and $(\phi_{1,2})_{\text{shear}}$ (**c**), for $\theta_1 = 60°$, $\theta_2 = 120°$, $\dot{\theta}_1 = \ddot{\theta}_1 \sim 0$, and $\dot{\theta}_2 = -1000$ and -2000 deg/s (intermediate phase of upper-limb push movement). In this condition, $\ddot{\theta}_2$ coincides with the elbow angular acceleration. F_x, F_y, and θ_F are reported for different values of elbow muscle torque τ_2 and negligible shoulder muscle torque ($\tau_1 = 0$). Only the selective effect of $\ddot{\theta}_2$ on the joint reaction forces was considered in Figure 2c.

The $(\phi_{0,1})_{\text{shear}}$ and $(\phi_{1,2})_{\text{axial}}$ components reach values as high as 300 N at the higher values of elbow extension angular velocity and acceleration (Figures 3c and 4c). This corresponds to an anterior shear force acting on the humerus and an axial forearm traction. The forearm also experiences an anterior shear force ($(\phi_{1,2})_{\text{shear}} > 0$), which is independent of $\ddot{\theta}_2$ and enhanced by $\left|\ddot{\theta}_2\right|$, and reaches a maximum magnitude of approximately 100 N. The elbow angular velocity and acceleration have opposite effects on the axial force $(\phi_{0,1})_{axial}$ acting on the humerus: $(\phi_{0,1})_{axial}$ is of traction type (positive) at high angular velocity and low angular acceleration of elbow extension and of compressive type (negative) at low angular velocity and high angular acceleration of elbow extension.

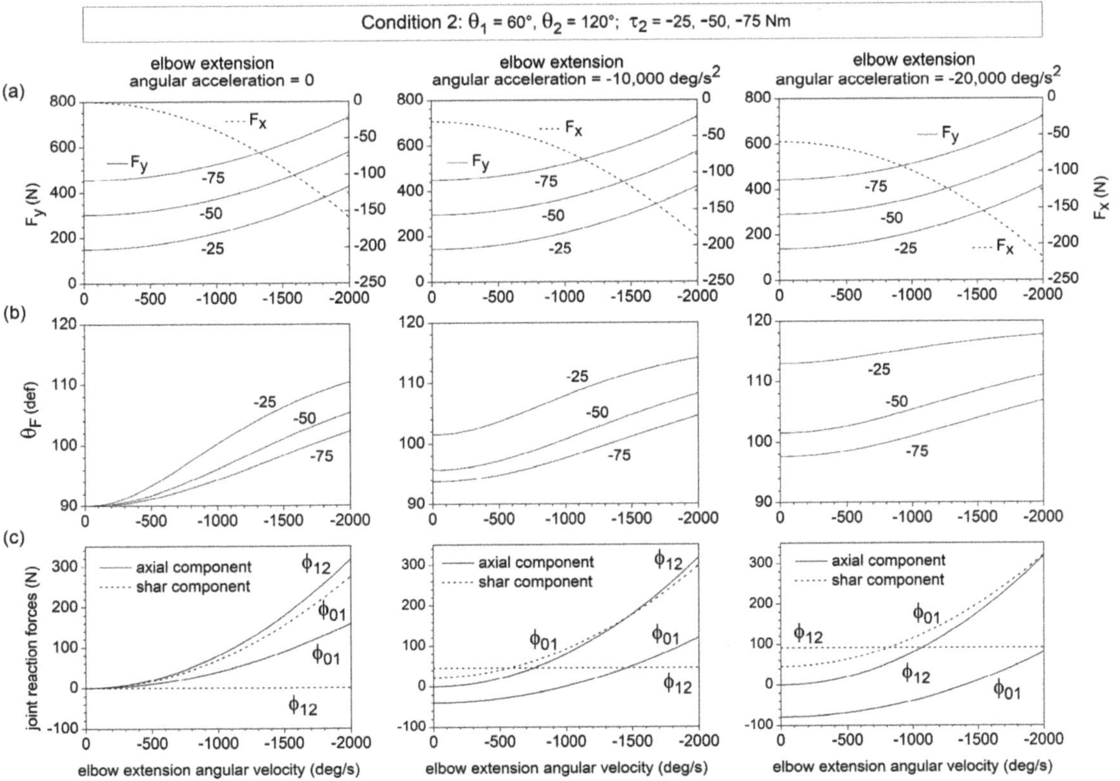

Figure 4. Dependence on the forearm angular velocity $\dot{\theta}_2$ of the components F_x and F_y of the endpoint force $\vec{F}$ (**a**), the direction θ_F of $\vec{F}$ (**b**), and the shear and axial components of the joint reaction forces $(\phi_{0,1})_{axial}$, $(\phi_{0,1})_{shear}$, $(\phi_{1,2})_{axial}$, and $(\phi_{1,2})_{shear}$ (**c**), for $\theta_1 = 60°$, $\theta_2 = 120°$, $\dot{\theta}_1 = \ddot{\theta}_1 \sim 0$, and $\ddot{\theta}_2 = 0, -10,000$, and $-20,000$ deg/s² (intermediate phase of upper-limb push movement). In this condition, $\dot{\theta}_2$ coincides with the elbow angular velocity. F_x, F_y, and θ_F are reported for different values of elbow muscle torque τ_2 and negligible shoulder muscle torque ($\tau_1 \sim 0$). Only the selective effect of $\dot{\theta}_2$ on the joint reaction forces was considered in Figure 2c.

4. Discussion

This study provides a dynamic model for a planar two-link chain controlled by single-joint and two-joint muscles. Equations (9)–(11) fully characterize the chain endpoint force $\vec{F}$ in relation to the chain configuration, the link angular velocities and accelerations, and the joint torques developed by the single-joint and two-joint muscles. Likewise, Equations (12)–(15) determine the axial and shear components of the joint reaction force acting on the joints, given the chain configuration, the link angular velocities and accelerations, and the forces of the single-joint and two-joint muscles. These equations were used to analyze upper-limb ballistic push movements against resistance in the transverse plane, focusing on the dynamic effects of the shoulder and elbow angular acceleration in the initial phase of the movement and of the forearm angular velocity and acceleration in the subsequent phase of the movement.

The results highlight that the shoulder and elbow angular velocity and acceleration influence the direction of the endpoint force $\vec{F}$. In the static condition, the shoulder muscle torque produces endpoint force directed along the forearm axis, while the elbow muscle torque produces endpoint force along the radial axis connecting the shoulder to the endpoint [4]. In both chain conditions examined (see Section 2), this corresponds to a static endpoint force $\vec{F}$ directed along the y-axis ($\theta_F = 90°$). The numerical simulation outcomes prove that a gradual increase in shoulder flexion angular acceleration in the initial phase of the movement, and in elbow extension angular velocity and acceleration in the subsequent phase of the movement, induce a progressive medial deviation in the $\vec{F}$ direction, with a shift in θ_F from 90° to 120° within the domain of interest of the values of the kinematic variables (Figures 2b, 3b and 4b). This information may have useful practical implications, as the optimization of the endpoint force direction constitutes a critical performance factor in pushing and throwing sport activities [25].

The link angular velocities and accelerations also significantly affect the value of specific axial and shear components of the joint reaction forces acting on the shoulder and elbow (Figures 2c, 3c and 4c). The shear force acting on the humerus (parallel to $\hat{w}_1$), produced by the humerus angular acceleration, is transmitted to the forearm through the elbow joint. This force acting on the forearm remains of the shear type (parallel to $\hat{w}_2$) if the elbow is extended, becomes axial type (parallel to $\hat{u}_2$) if the elbow is flexed at 90°, and gives rise to both axial and sheer forearm forces when the elbow is at 120° of flexion (as in the examined condition 1), with a prevalence of the axial component (Figure 2c). This simple example points out that the chain configuration plays a critical role in the transmission of axial and shear joint forces along the chain. Most importantly, the combined contribution of the elbow extension angular velocity and acceleration may give rise to shoulder shear forces and elbow traction forces as high as 300 N (Figure 4c). Advanced 3D models that consider the periarticular soft tissues and the complex geometry of the articular surfaces of the shoulder and elbow are necessary to estimate how the axial and shear joint forces are distributed among the different joint structures.

Equations (9)–(11) express the endpoint force components F_x and F_y as a function of the joint torques τ_1 and τ_2 and the chain kinematics (θ_1, θ_2, $\dot{\theta}_1, \dot{\theta}_2, \ddot{\theta}_1$, and $\ddot{\theta}_2$). However, these equations can also be used to determine τ_1 and τ_2 from the values of the variables θ_1, θ_2, $\dot{\theta}_1, \dot{\theta}_2, \ddot{\theta}_1, \ddot{\theta}_2$, F_x, and F_y measured experimentally, for example, by a motion analysis system and load cells. This, in conjunction with electromyographic (EMG) measurements, would provide useful information on muscle function during two-joint exercises involving fast movements and external resistance.

The main limitation of this study concerns the kinematic conditions of the chain used as input parameters in the model, as defined by Equations (9)–(15). The values of these input parameters were not measured directly but selected according to the results of previous experimental studies reporting the kinematic patterns recorded during upper limb ballistic movements [8–12]. Overall, these previous studies examined groups of high-level athletes involved in different sports activities that were performed in specific sports environments. For this reason, we believe that the selected conditions refer to actual critical phases of upper limb ballistic movements that are of particular interest in sports and exercise science. Importantly, the explicit analytical formulation of the general Equations (9)–(15) enables the analysis of any other chain condition of specific interest.

Another limitation stems from the analysis of the shear and axial components of the joint reaction forces (Figures 2–4). Only the contribution of the link angular velocities and accelerations was analyzed. The contribution of the resistance $\vec{R}$ ($\vec{R} = -\vec{F}$) can be immediately derived from Equations (12)–(15). Conversely, the determination of the contribution of the force of each single muscle in the chain would require knowledge of the muscle line of action in the specific chain configuration. The real magnitude of a muscle force can be estimated from the muscle architecture parameters and the degree of muscle activation with the use of EMG-driven musculoskeletal models and optimization procedures [26–30]. However, this is far beyond the scope of the present study. Nevertheless, for any set of values for the line of action and magnitude of the muscle forces, Equations (12)–(15) determine the contribution of the force of each muscle to the shear and axial components of the reaction force acting on each joint in the chain.

The intersegment dynamic model defined by Equations (9)–(15) was applied to upper limb ballistic movements in the transverse plane. Notably, however, the model can also be applied to any planar upper or lower limb movement by simply considering the components of the weights of the links in the plane of movement. Furthermore, the contribution of the joint angular velocity and acceleration to the axial and shear components of the joint reaction forces (Figures 2c, 3c and 4c) does not depend on the plane of movement. Therefore, the results of the present study can be used to quantify the axial and shear joint loads during several athletic and sport ballistic actions, including kicking a ball, punching, shot putting, explosive bench press throw, medicine ball throws, and overhead throwing.

5. Conclusions

The dynamic Equations (9)–(15) provide the closed-form analytical expressions of the endpoint force, and the shear and axial components of the joint reaction forces, for a planar two-link musculoskeletal chain controlled by single-joint and two-joint muscles. Numerical simulations highlight that the shoulder and elbow angular velocities and accelerations typically reached during ballistic sport movements can considerably affect the endpoint force direction and the shear and axial joint loads. This information can help estimate the mechanical stress acting on specific joint structures during explosive/ballistic sport movements and power-oriented exercises. It also helps in optimizing the endpoint force direction to improve the performance during these activities.

Funding: This research received no external funding.

Institutional Review Board Statement: Not applicable.

Informed Consent Statement: Not applicable.

Data Availability Statement: The data that support the findings of this study are available from the corresponding author upon request.

Acknowledgments: The author thanks Samuele Contemori for technical support.

Conflicts of Interest: The author declares no conflict of interest.

References

1. Craig, J.J. *Introduction to Robotics: Mechanics and Control*, 3rd ed.; Pearson Prentice Hall: Upper Saddle River, NJ, USA, 2004.
2. Siciliano, B.; Sciavicco, L.; Villani, L.; Oriolo, G. *Robotics Modelling, Planning and Control*; Springer: London, UK, 2009.
3. Zatsiorsky, V.M. *Kinematics of Human Motion*; Human Kinetics: Champaign, IL, USA, 1998.
4. Zatsiorsky, V.M. *Kinetics of Human Motion*; Human Kinetics: Champaign, IL, USA, 2002.
5. Hollerbach, J.M.; Flash, T. Dynamic interactions between limb segments during planar arm movements. *Biol. Cybern.* **1982**, *44*, 67–77. [CrossRef]
6. Sainburg, R.L.; Ghilardi, M.F.; Poizner, H.; Ghez, C. Control of limb dynamics in normal subjects and patients without proprioception. *J. Neurophysiol.* **1995**, *73*, 820–835. [CrossRef] [PubMed]
7. Latash, M.L.; Zatsiorsky, V.M. *Biomechanics and Motor Control. Defining Central Concepts*; Academic Press: London, UK, 2016.
8. Wagner, H.; Pfusterschmied, J.; Tilp, M.; Landlinger, J.; von Duvillard, S.P.; Müller, E. Upper-body kinematics in team-handball throw, tennis serve and volleyball spike. *Scand. J. Med. Sci. Sports* **2014**, *24*, 345–354. [CrossRef]
9. Escamilla, R.F.; Barrentine, S.W.; Fleisig, G.S.; Zheng, N.; Takada, Y.; Kingsley, D.; Andrews, J.R. Pitching biomechanics as a pitcher approaches muscular fatigue during a simulated baseball game. *Am. J. Sports Med.* **2007**, *35*, 23–33. [CrossRef]
10. Matsuo, T.; Escamilla, R.F.; Fleisig, G.S.; Barrentine, S.W.; Andrews, J.R. Comparison of kinematic and temporal parameters between different pitch velocity groups. *J. Appl. Biomech.* **2001**, *17*, 1–13. [CrossRef]
11. Werner, S.L.; Suri, M.; Guido, J.A.; Meister, K.; Jones, D.G. Relationships between ball velocity and throwing mechanics in collegiate baseball pitchers. *J. Shoulder Elbow Surg.* **2008**, *17*, 905–908. [CrossRef]

12. Kaizu, Y.; Watanabe, H.; Yamaji, T. Correlation of upper limb joint load with simultaneous throwing mechanics including acceleration parameters in amateur baseball pitchers. *J. Phys. Ther. Sci.* **2018**, *30*, 223–230. [CrossRef]
13. Sakamoto, A.; Kuroda, A.; Sinclair, P.J.; Naito, H.; Sakuma, K. The effectiveness of bench press training with or without throws on strength and shot put distance of competitive university athletes. *Eur. J. Appl. Physiol.* **2018**, *118*, 1821–1830. [CrossRef]
14. Biscarini, A. Non-Slender n-Link Chain Driven by Single-Joint and Multi-Joint Muscle Actuators: Closed-Form Dynamic Equations and Joint Reaction Forces. *Appl. Sci.* **2021**, *11*, 6860. [CrossRef]
15. Escamilla, R.F.; Fleisig, G.S.; Zheng, N.; Barrentine, S.W.; Wilk, K.E.; Andrews, J.R. Biomechanics of the knee during closed kinetic chain and open kinetic chain exercises. *Med. Sci. Sports Exerc.* **1998**, *30*, 556–569. [CrossRef] [PubMed]
16. Shelburne, K.B.; Torry, M.R.; Pandy, M.G. Muscle, Ligament, and Joint-Contact Forces at the Knee during Walking. *Med. Sci. Sports Exerc.* **2005**, *37*, 1948–1956. [CrossRef] [PubMed]
17. Biscarini, A. Minimization of the knee shear joint load in leg-extension equipment. *Med. Eng. Phys.* **2008**, *30*, 1032–1041. [CrossRef] [PubMed]
18. Yanagawa, T.; Goodwin, C.J.; Shelburne, K.B.; Giphart, J.E.; Torry, M.R.; Pandy, M.G. Contributions of the individual muscles of the shoulder to glenohumeral joint stability during abduction. *J. Biomech. Eng.* **2008**, *130*, 021024. [CrossRef]
19. Reinold, M.M.; Escamilla, R.F.; Wilk, K.E. Current concepts in the scientific and clinical rationale behind exercises for glenohumeral and scapulothoracic musculature. *J. Orthop. Sports Phys. Ther.* **2009**, *39*, 105–117. [CrossRef] [PubMed]
20. Biscarini, A.; Benvenuti, P.; Botti, F.M.; Mastrandrea, F.; Zanuso, S. Modelling the joint torques and loadings during squatting at the Smith machine. *J. Sports Sci.* **2011**, *29*, 457–469. [CrossRef]
21. Biscarini, A. Determination and optimization of joint torques and joint reaction forces in therapeutic exercises with elastic resistance. *Med. Eng. Phys.* **2011**, *34*, 9–16. [CrossRef] [PubMed]
22. Biscarini, A.; Contemori, S.; Busti, D.; Botti, F.M.; Pettorossi, V.E. Knee flexion with quadriceps cocontraction: A new therapeutic exercise for the early stage of ACL rehabilitation. *J. Biomech.* **2016**, *49*, 3855–3860. [CrossRef]
23. Khoddam-Khorasani, P.; Arjmand, N.; Shirazi-Adl, A. Effect of changes in the lumbar posture in lifting on trunk muscle and spinal loads: A combined in vivo, musculoskeletal, and finite element model study. *J. Biomech.* **2020**, *104*, 109728. [CrossRef]
24. Winter, A. *Biomechanics and Motor Control of Human Movement*, 4th ed.; Wiley: Hoboken, NJ, USA, 2009; p. 109.
25. Zatsiorsky, V.M. *Biomechanics in Sport*; Blackwell Science: London, UK, 2000.
26. Pandy, M.G. Computer modeling and simulation of human movement. *Annu. Rev. Biomed. Eng.* **2001**, *3*, 245–273. [CrossRef]
27. Prilutsky, B.I.; Zatsiorsky, V.M. Optimization-Based Models of Muscle Coordination. *Exerc. Sport Sci. Rev.* **2002**, *30*, 32–38. [CrossRef]
28. Erdemir, A.; McLean, S.; Herzog, W.; Van den Bogert, A. Model-based estimation of muscle forces exerted during movements. *Clin. Biomech.* **2007**, *22*, 131–154. [CrossRef] [PubMed]
29. Gagnon, D.; Arjmand, N.; Plamondon, A.; Shirazi-Adl, A.; Larivière, C. An improved multi-joint EMG-assisted optimization approach to estimate joint and muscle forces in a musculoskeletal model of the lumbar spine. *J. Biomech.* **2011**, *44*, 1521–1529. [CrossRef] [PubMed]
30. Sharifzadeh-Kermani, A.; Arjmand, N.; Vossoughi, G.; Shirazi-Adl, A.; Patwardhan, A.G.; Parnianpour, M.; Khalaf, K. Estimation of Trunk Muscle Forces Using a Bio-Inspired Control Strategy Implemented in a Neuro-Osteo-Ligamentous Finite Element Model of the Lumbar Spine. *Front. Bioeng. Biotechnol.* **2020**, *8*, 949. [CrossRef] [PubMed]

bioengineering

Article

The Effect of Fatigue on Postural Control and Biomechanical Characteristic of Lunge in Badminton Players

Yanyan Du [1,2,*] and Yubo Fan [1,*]

1   Key Laboratory for Biomechanics and Mechanobiology of Ministry of Education, Beijing Advanced Innovation Center for Biomedical Engineering, School of Biological Science and Medical Engineering, and with the School of Engineering Medicine, Beihang University, Beijing 100083, China

2   Beijing Key Laboratory of Sports Function Assessment and Technical Analysis, School of Kinesiology and Health, Capital University of Physical Education and Sports, Beijing 100191, China

*   Correspondence: duyanyan_dy@126.com (Y.D.); yubofan@buaa.edu.cn (Y.F.); Tel.: +86-10-8233-9428 (Y.F.)

Abstract: This study investigated the effects of fatigue on postural control and biomechanical characteristic of lunge. A total of twelve healthy male collegiate badminton players (21.1 ± 2.2 years; 180.8 ± 4.0 cm; 72.5 ± 8.4 kg; 8.9 ± 3.5 years of experience) performed repeating lunges until exhausted. Postural stability was evaluated through a single-leg balance test using the dominant lower limb on a pressure plate with eyes opened (EO) and eyes closed (EC). The center of pressure (CoP) sway in the entire plantar and sub-regions of the plantar was measured. Kinematic and kinetic data of lunge motion were collected. The postural control was impaired after fatigue. In plantar sub-regions, the area, displacement and distance in the medial–lateral (ML) and anterior–posterior directions of CoP increased significantly ($p < 0.05$), especially the distance in ML. The medial region of the forefoot is the most sensitive to fatigue. Compared to pre-fatigue, participants experienced a significantly longer phase of pre-drive-off ($p < 0.01$), less peak moment and peak power of the knee and hip for drive-off ($p < 0.01$) and less peak moment of the ankle during braking phase ($p < 0.05$). These findings indicate that, within the setting of this investigation, the different responses to fatigue for CoP sway in plantar sub-regions and the consistency between postural control and biomechanical characteristic of lunge may be beneficial for developing and monitoring a training plan.

Keywords: balance; COP; fatigue; badminton; kinematics; kinetics

Citation: Du, Y.; Fan, Y. The Effect of Fatigue on Postural Control and Biomechanical Characteristic of Lunge in Badminton Players. *Bioengineering* **2023**, *10*, 301. https://doi.org/10.3390/bioengineering10030301

Academic Editor: Aurélien Courvoisier

Received: 25 November 2022
Revised: 14 February 2023
Accepted: 17 February 2023
Published: 27 February 2023

1. Introduction

In badminton, balance is one of the key factors for an effective shot. Previous studies have demonstrated that muscle fatigue can reduce the maximum voluntary muscle force and work capacity [1] and impair the proprioceptive system, which in turn can contribute to a decrease in joint stability, and can alter movement control [2] and impair the postural control [3–5]. Additionally, these performance changes under fatigue might increase the risk of injury [6]. An epidemiological study reported that badminton is one of the games with a high injury rate [7]. Overuse injury caused by high-intensity training and repetitive movement is the main type of injury [8,9]. However, the responses of fatigue on balance in badminton are unknown, and information is needed to provide insight about the postural control changes which occur after fatigue.

A wide variety of fatigue protocols were used in fatigue studies, including isometric [5], treadmill running, specific fatigue [3,10], simulating match [11] and so forth. These are generally categorized as either general fatigue or peripheral nerve fatigue. Additionally, it is verified that the fatigue protocols did not uniformly produce alterations in lower-limb neuromuscular factors related to the high risk of injuries, such as an ACL injury [12]. Density and duration of training [13], anatomical position to move [14] and duration of concentric/eccentric movement would all lead to distinct muscle activation and neuromuscular fatigue responses. Consequently, fatigue protocol is the key part for the interpretation of results. Considering that fatigue effects occur cumulatively throughout a practice or a game and the entire athletic season [12,15], some protocols attempt to simulate realistic match [11] or game situations focusing on a specific relevant skill [16] to provide support for training. In badminton, the lunging action is a basic and important footwork [17]. When repeating lunges until fatigued, the activity of vastus lateralis and biceps femoris showed significant change [18]. Joint stiffness in the knees during a forward lunge task increased after fatigue due to the repeated forward lunges by badminton players. This would then influence their performance and increase the injury risk [19]. A study has reported that the dominant (leading) limb, acting as a generator of vertical force during a lunge, may contribute to the development of muscular imbalances, which may ultimately contribute to the development of an overuse injury [20]. Compared to other directions, the left-forward and right-forward lunge directions were associated with higher plantar loading at the heel and toe regions. However, few studies pay attention to the changes of postural control and lunge performance after fatigue in badminton.

Generally, the sway of center of pressure (CoP) standing on a force plate system [21] or balance system [22] is used to assess the postural sway. However, the pressure plate system was confirmed to provide reliable and valid measures of static standing balance [9]. In addition, significant differences were found in the pressure distribution of plantar sub-regions [15,23,24] during the stance phase of a lunge. These differences in the postural, caused by fatigue, may also induce a different distribution of plantar pressure. However, to the best of the authors' knowledge, there is no research that has studied and discussed this.

Consequently, the aim of the present study was to investigate the effect of fatigue caused by repeating lunges on the postural control and the biomechanical characteristics of lunges. To reach the objective, repeating forward lunges until exhaustion was proposed as fatigue protocol. The pressure plate was used to access the sway of CoP in whole and sub-regions of the dominant plantar. Additionally, the following hypotheses were formulated: (1) postural sway would increase after the fatigue protocol, (2) different sub-regions of the plantar would have different responses to fatigue and (3) the kinematics and kinetics of the lunges would be affected by fatigue.

2. Materials and Methods

A total of twelve healthy male collegiate badminton athletes (21.1 ± 2.2 years; 180.8 ± 4.0 cm; 72.5 ± 8.4 kg; 8.9 ± 3.5 years of experience; 2–4 h badminton training per day) were recruited by badminton coaches from two universities for the study. All participants were free from musculoskeletal or neurological conditions, and without lower limb injuries in the last three months.

Prior to participating in the study, all participants were informed of the experimental procedures and potential risks, and then they signed an informed consent document. Additionally, a questionnaire about their anthropometrics, healthy status, injuries history, training plan and physical activity levels were filled out. They all wore badminton footwear of the same brand and series, avoiding the effect of footwear. The experiment was approved by the Ethics committee of Beihang University.

Prior to the test, participants took 15 min to warm up and become familiar with the footwork and court. In addition, they were sure to stop when they experienced discomfort at any time. Then, we started the test. Heart rate (HR), blood lactate (BL) and a Borg 6–20 rating of perceived exertion (RPE) were assessed at baseline (0), immediately (T0) after exhaustion and at the 6th (T6) and the 9th minute (T9) after the protocol, accounting for the time for the testing of HR, BL and RPE. Participants completed the single-leg balance test at baseline and immediately following the HR, BL and RPE testing after the fatigue protocol. Kinematics and kinetics of the lunge were recorded by motion capture system and force platform. After fatigue, all tests were completed within 10 min.

Repeating forward lunge until exhaustion was proposed as the fatigue protocol [25]. The protocol was elaborated to be conducted in a simulated badminton court. One forward lunge was defined as starting from the starting position, lunging with a sliding step to the force platform, kicking the shuttlecock, and moving back to the starting position (details are illustrated in part I of Figure 1). An 80–90% maximus range of lunge was repeated once for approximately 2.8 s until they were exhausted. A metronome was used to establish the rate. Participants were instructed to follow the rhythm, and stop when they could not follow the rhythm for 10 s.

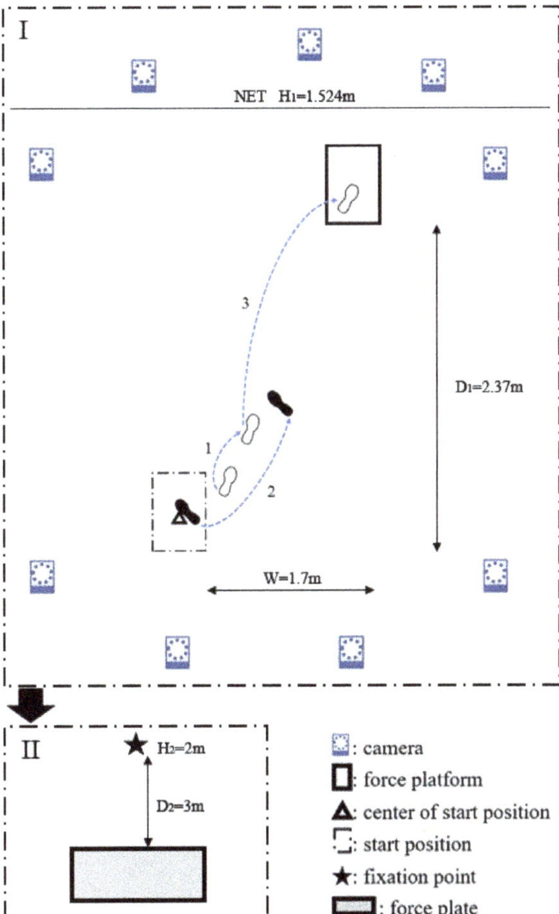

Figure 1. Illustration of footwork and testing process. The badminton athletes lunged and followed the footwork shown in I, and repeated until exhausted. Then, they took II, the poster sway test, standing on a plantar pressure plate using their dominant leg. In the illustration of lunge footwork, the right leg is dominant. The open foot marks represent the foot placements of the right foot, whereas the solid foot marks represent the foot placements of the left foot. The numbers represent the step sequences.

HR (beats per minute, bpm) was assessed by the Polar heart rate sensor H1 (Polar Electro, Kempele, Finland) throughout the test processing. BL (mmol/L) was measured by a portable blood lactate meter (SensLab GmbH, h/p/cosmos sirius®, Leipzig, Germany). Additionally, RPE was recorded. Fatigue was decisively judged by the values of maximum HR [18], BL and RPE [26].

Here, maximum HR (HRmax) was calculated as 220 minus age (year). The criteria included (1) HR being greater than or equal to HRmax, (2) BL being greater than or equal to 8 mmol/L and (3) the value of RPE being greater than or equal to 18. Once two of the criteria were fitted, the participants were exhausted. The testing was completed, and it was a valid test. Additionally, the testing was interrupted when one participant was too fatigued to continue the testing, even if no criterion was reached.

Participants stood upright on a pressure plate (40 × 100 cm, 1 m-3D; footscan® system, RSscan International, Olen, Belgium) in a one-leg posture with the dominant limb, arms by their side. They were instructed to stand as quietly as they could with eyes opened (EO), looking at a target positioned 3 m away (details are illustrated in part II of Figure 1), and with eyes closed (EC). Participants lifted the non-dominant leg upon an auditory signal. A primary investigator performed all balance tests. The balance module of pressure plate system was set with a 5 s delay after the auditory signal and to record for 5 s (100 Hz, 128 Lines/plate).

Plantar was divided into top (forefoot) and bottom (rearfoot) regions (T, B), and then further divided into medial and lateral regions (TM, TL, BM, BL, illustrated in Figure 2), using the software of pressure plate system. Threshold level for pressure was 10 N. The center of pressure (CoP) sway of entire and sub-regions were exported from the software. For each trial, the following variables were calculated: ellipse area containing 95% of the CoP data points (Area), sway displacement (Dis) and distance in medial–lateral (ML) and anterior–posterior (AP) regions.

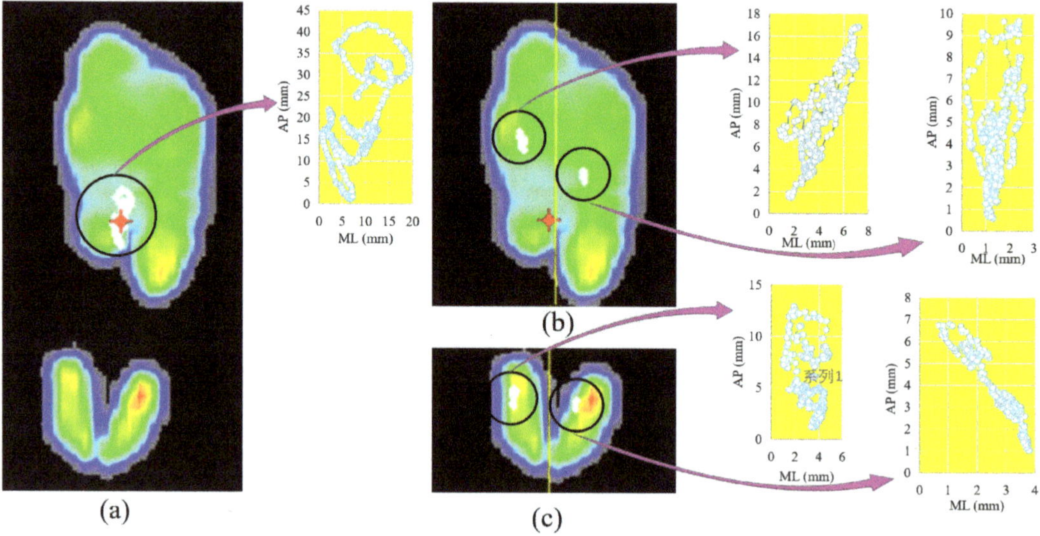

Figure 2. One example of the center of pressure (CoP) of planter and its sub-regions while standing on the dominant leg with eyes opened, before fatigue. (**a**) The maximum-pressure image and the sway of CoP; (**b**) the top of plantar is divided into two sub-regions along the yellow line, and the sway of CoP; (**c**) the bottom of plantar is divided to two sub-regions along the yellow line, and the sway of CoP of plantar and sub-regions.

A nine-camera motion capture system (Qualysis, Göteborg, Sweden) sampling at 200 Hz and one Kistler force platform (Kistler, 9286 A, Kistler Instrument AG, Winterthur, Switzerland) sampling at 1000 Hz were used to collect the ground reaction forces (GRF) and kinematic data, simultaneously. According to the CAST lower-leg model [27], reflective markers (18 mm diameter) were firmly placed over the hip and lower legs. The kinematic and kinetic data of lunge motion were collected.

The kinematic and force data were obtained by the optical motion capture system and then exported and saved as c3d files. Then, for the dominant lower limb, the hip, knee, and ankle joint angles, moments, power and ground reaction force (GRF) during stance phase of lunge were calculated using visual 3D software (V5, C-Motion, Bethesda, MD, USA). The raw kinematic data were filtered with a low-pass (Butterworth) filter, with a frequency of 20 Hz [10]. The threshold of the vertical ground reaction force (vGRF) data was set at 10 N.

The stance phase, from initial contact (heel strike) to final lift-off from the force-plate by the dominant limb, was determined by the vGRF value. Based on previous studies [20,25], the stance phase was divided into five phases: I (0~initial impact peak (PF1)), II (PF1~secondary impact peak (PF2)), III (PF2~peak angle of knee flexion (PAK)), IV (PAK~third peak during drive-off (PF3)) and V(PF3~end). Based on the previous literature linked to the lunge in badminton [15,20,25], we analyzed the impact peak, duration of five sub-stance phases, hip, knee, and ankle joint initial contact angles, durations to peak angle, ranges of motion (RoM), peak angles, moments and power in the sagittal plane.

Results are reported as mean $\pm$ standard deviation (SD). All variables in this study were examined for normality using a Shapiro–Wilk test prior to statistics analysis. The force, joint moment and power were normalized to body mass. One-way Repeated-Measures ANOVA was used for the analysis of the influence of fatigue on the related parameters, including fatigue parameters (HR, BL and RPE) between the baseline (0), T0, T6 and T9, the postural sway variables and biomechanical data. Paired t-tests were performed to identify differences. All statistical procedures were performed with IBM SPSS Statistics for Window (Version 25.0; IBM Corp., NY, USA). Additionally, a statistical significance level was accepted at 0.05. Effect size (Cohen's d) was computed for the t-test. Small, middle and large effect sizes were $0.2 \leq d < 0.5$, $0.5 \leq d < 0.8$ and $d > 0.8$, respectively.

3. Results

For the HR, BL and RPE, the mean and SD values are illustrated in Figure 3. Significant increases were observed for HR ($p < 0.001$), BL ($p < 0.001$) and RPE ($p < 0.001$). Although these values decreased at T6 and T9, they showed significant differences from the baseline ($p < 0.01$).

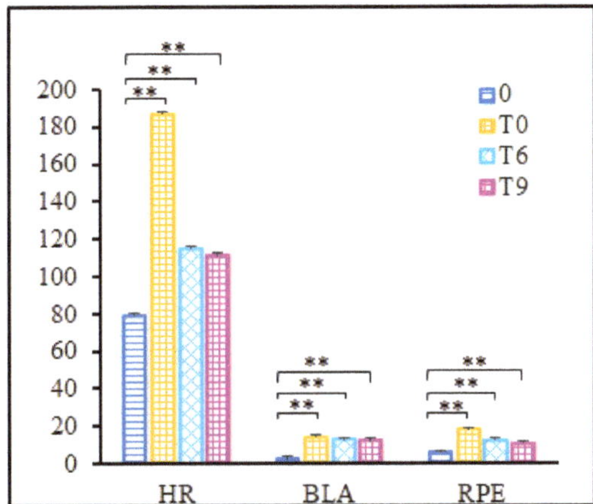

Figure 3. Rate of heart (HR) (beats per minute, bpm), blood lactate (BLA) (mmol/L) and Borg 6–20 rating of perceived exertion (RPE). 0—baseline, before fatigue protocol; T0—immediately after fatigue protocol; T6—the 6th minute after fatigue protocol; T9—the 9th minute after fatigue protocol. Note: ** indicates the significance level $p < 0.001$.

For the variables of postural sway in the entire region, Figure 4 (Entirety) shows that with either eyes opened or closed, all variables had no significant change after the fatigue protocol.

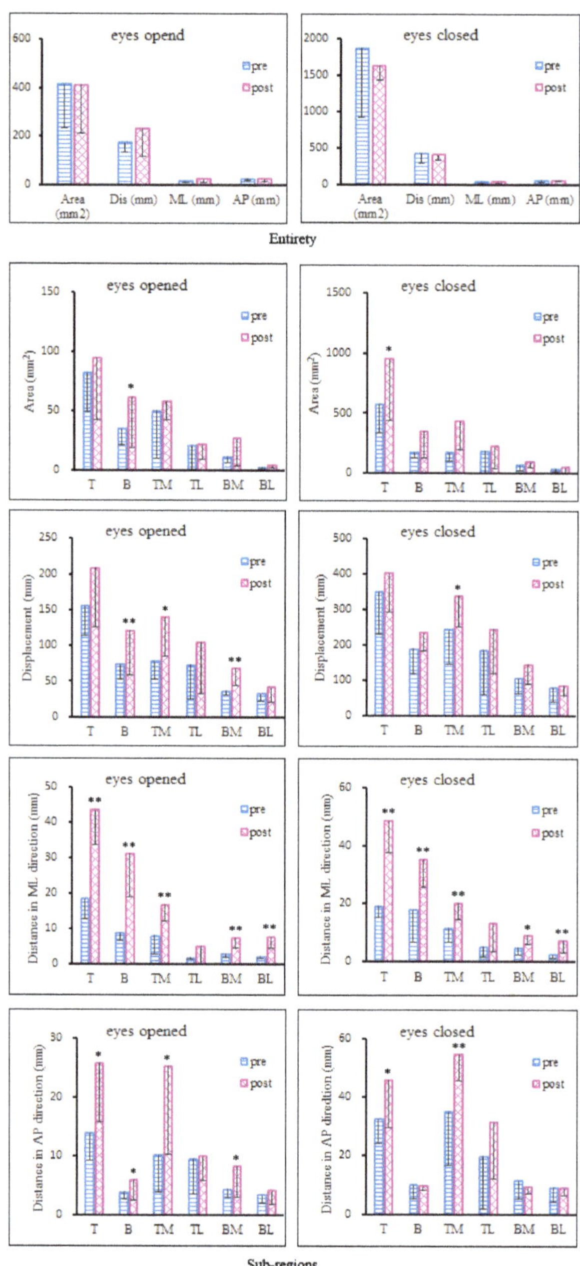

Figure 4. Standing on the dominant leg with eyes opened and closed. Plantar was divided into the following six sub-regions: top (T, forefoot), bottom (B, rearfoot), medial and lateral of the top (TM, TL) and medial and lateral of the bottom (BM, BL). The area (mm^2), displacement (mm), sway distance in medial–lateral (ML) and anterior–posterior (AP) regions of postural sway are shown for the entirety and sub-regions using the pre- and post-fatigue protocols. * indicates the significance level $p < 0.05$; ** indicates the significance level $p < 0.01$.

For sub-regions, both with EO and EC, almost all of the postural variables for all sub-regions increased after the fatigue protocol (illustrated in Figure 4, Sub-regions). With EO, area of rearfoot ($p = 0.04$), displacement of rearfoot ($p = 0.006$), medial forefoot ($p = 0.046$), medial rearfoot ($p = 0.003$), distance in ML of forefoot ($p < 0.001$), rearfoot ($p < 0.001$), medial forefoot ($p = 0.001$), medial rearfoot ($p < 0.001$), lateral rearfoot ($p < 0.001$), distance in AP of forefoot ($p = 0.014$), rearfoot ($p = 0.035$), medial forefoot ($p = 0.031$) and medial rearfoot ($p = 0.032$) increased significantly within 10 min after the protocol (Figure 4). With EC, variables also increased significantly, specifically, area of forefoot ($p = 0.05$), displacement of medial forefoot ($p = 0.035$), distance in ML of forefoot ($p < 0.001$), rearfoot ($p < 0.001$), medial forefoot ($p = 0.002$), lateral forefoot ($p = 0.017$), medial rearfoot ($p = 0.014$), lateral rearfoot ($p = 0.006$), distance in AP of forefoot ($p = 0.029$) and medial forefoot ($p = 0.004$) (Figure 4).

Table 1 shows that the stance phase was divided into five sub-phases by the time of peak force (initial impact, secondary impact and drive-off) and peak angle (ankle plantar flexion and knee flexion). For pre- and post-fatigue, there were statistically significant differences in the time of drive-off impact peak ($p = 0.005$, $d = 1.410$); the duration of phases III, IV and V ($p = 0.041, 0.002, 0.005$; $d = 0.883, 1.701, 1.410$, respectively); time of peak angle (T%) for ankle and knee in sagittal plane ($p = 0.001, 0.009$; $d = 1.972, 1.270$, respectively).

Table 1. Time of peak force (T%), sub-phases (T%) and time of peak angle (T%) (mean ± SD).

	Pre-Fatigue	Post-Fatigue	t	p	Cohen's d
Time of peak force (T%)					
Initial impact	3.9 (0.6)	3.8 (0.7)	0.4	0.685	0.150
Secondary impact	15 (3.2)	12.6 (3.3)	2.2	0.065	0.774
Drive-off impact	73.1 (5.6)	77 (6.7)	−4	0.005 *	1.410
Phases (T%)					
I (0-PF1)	3.9 (0.6)	3.8 (0.7)	0.4	0.685	0.150
II (PF1-PF2)	11.1 (2.7)	8.9 (3.3)	1.8	0.108	0.651
III (PF2-PAK)	30.9 (6)	21 (9.3)	2.5	0.041 *	0.883
IV (PAK-PF3)	27.3 (10.7)	43.4 (15.2)	−4.8	0.002 *	1.701
V (PF3-end)	26.9 (5.6)	23 (6.7)	4	0.005 *	1.410
Time of peak angle (T%)					
ankle-sagittal	14.1 (2.9)	11.3 (2.7)	5.6	0.001 *	1.972
knee-sagittal	45.9 (6.5)	33.6 (9.4)	3.6	0.009 *	1.270
hip-sagittal	41.6 (6.6)	36.8 (6.9)	2	0.086	0.707

Notes: sagittal plane represents the flexion/extension (knee, hip) and dorsiflexion/plantar flexion (ankle). * indicates the significance level $p < 0.05$. Effect size (Cohen's d), small: $0.2 \leq d < 0.5$, middle: $0.5 \leq d < 0.8$ and large: $d > 0.8$.

The initial contact angle of the knee and the hip, the peak knee flexion and the knee RoM in sagittal plane decreased significantly after fatigue ($p = 0.004, 0.044, 0.008, 0.048$; $d = 1.476, 0.868, 1.296, 0.848$, respectively) (Table 2).

For PF1, PF2 and PF3, Table 3 shows no differences between pre- and post-fatigue. Peak joint moment and power in sub-phases were calculated. The peak joint moment in the sagittal plane (knee—I, II and III: $p = 0.004$, $d = 1.394$, IV and V: $p = 0.011$, $d = 1.445$; hip—I, II and III: $p = 0.027$, $d = 0.893$, IV and V: $p = 0.009$, $d = 1.230$) and in the frontal plane (ankle—II: $p = 0.026$, $d = 0.996$, III: $p = 0.033$, $d = 0.934$), and joint power in the sagittal plane (knee—I, II and III: $p = 0.009$, $d = 1.268$, IV and V: $p = 0.000$, $d = 2.325$; hip—IV and V: $p = 0.014$, $d = 1.157$) decreased after fatigue.

Table 2. Initial contact and peak angle, and range of motion (RoM) of ankle, knee and hip (mean ± SD).

	Pre-Fatigue	Post-Fatigue	t	p	Cohen's d
Initial contact angle (degree)					
ankle-sagittal	10.4 (8.0)	6.4 (9.1)	1.4	0.212	0.486
knee-sagittal	13.1 (7.4)	7.9 (6.1)	4.2	0.004 *	1.476
hip-sagittal	46.1 (12.7)	42.2 (11.8)	2.5	0.044 *	0.868
ankle-frontal	11.1 (7.5)	13.4 (7.2)	−1.5	0.180	0.527
knee-horizontal	−19.1 (10.7)	−28.3 (19.1)	1.9	0.106	0.657
Peak angle (degree)					
ankle-sagittal (II)	−20.8 (5.7)	−18.2 (6.5)	−1.9	0.098	0.674
knee-sagittal	69.5 (8.8)	62.5 (8.8)	3.7	0.008 *	1.296
hip-sagittal	74.1 (11.9)	67.8 (12.8)	1.5	0.189	0.514
ankle-frontal (III and IV)	−3.4 (5.9)	−1.6 (4.5)	−1.1	0.337	1.395
knee-horizontal	1.4 (5.6)	−7 (17.3)	1.5	0.178	0.529
Range of motion (degree)					
ankle-sagittal	34.1 (8.1)	32 (8.2)	0.6	0.563	0.214
knee-sagittal	59.3 (7.3)	55.1 (6.3)	2.4	0.048 *	0.848
hip-sagittal	44.6 (10.4)	38.7 (10.8)	0.9	0.379	0.332
ankle-frontal	20.8 (9.8)	18.3 (3.5)	0.7	0.523	0.313
knee-horizontal	1.4 (5.6)	−6.9 (17.4)	1.5	0.180	0.526

Notes: sagittal plane represents the flexion/extension (knee, hip) and dorsiflexion/plantar flexion (ankle); frontal plane represents the eversion/inversion (ankle); horizontal plane represents the internal/external (knee). * indicates the significance level $p < 0.05$. Effect size (Cohen's d), small: $0.2 \leq d < 0.5$, middle: $0.5 \leq d < 0.8$ and large: $d > 0.8$.

Table 3. Peak force (N/BW), peak moment (Nm/BW) and power (W/BW) (mean ± SD).

	Pre-Fatigue	Post-Fatigue	t	p	Cohen's d
Peak force (N/BW)					
Initial	1.3 (0.1)	1.3 (0.2)	1.3	0.249	0.445
Secondary	1.6 (0.1)	1.5 (0.2)	1.7	0.142	0.584
Drive-off	1.3 (0.1)	1.2 (0.2)	1.6	0.158	0.558
Peak moment (Nm/BW)					
ankle-sagittal (I and II)	−0.6 (0.1)	−0.5 (0.1)	−2.1	0.072	0.679
ankle-sagittal (III and IV)	0.8 (0.1)	0.9 (0.3)	−0.9	0.402	0.341
knee-sagittal (I, II and III)	3 (0.7)	2.7 (0.5)	4.2	0.004 *	1.394
knee-sagittal (IV and V)	2.5 (0.6)	2 (0.5)	3.4	0.011 *	1.445
hip-sagittal (I, II andIII)	2.3 (0.5)	1.8 (0.6)	2.8	0.027 *	0.893
hip-sagittal (IV and V)	2.5 (0.4)	2.1 (0.5)	3.6	0.009 *	1.230
ankle-frontal (I)	0.1 (0.1)	0.1 (0.1)	−0.5	0.608	0.19
ankle-frontal (II)	−0.2 (0.2)	−0.1 (0.2)	−2.8	0.026 *	0.996
ankle-frontal (III)	−0.2 (0.2)	−0.1 (0.1)	−2.6	0.033 *	0.934
knee-horizontal (III and IV)	0.5 (0.5)	0.4 (0.4)	2.1	0.077	0.732
Peak power (W/BW)					
ankle-sagittal (I and II)	−6.1 (1.4)	−4.4 (1.8)	−1.9	0.104	0.661
knee-sagittal (I, II and III)	−17.2 (4.4)	−13.7 (3.1)	−3.6	0.009 *	1.268
knee-sagittal (IV and V)	10.9 (3.1)	6.1 (2.3)	6.6	0.000 *	2.325
hip-sagittal (I, IIandIII)	−6.9 (2.4)	−5.8 (2.5)	−1.6	0.15	0.571
hip-sagittal (IV and V)	4.4 (2.5)	2.2 (0.9)	3.3	0.014 *	1.157
knee-horizontal (I and II)	1.2 (0.8)	1.4 (0.8)	−0.8	0.474	0.288

Notes: sagittal plane represents the flexion/extension (knee, hip) and dorsiflexion/plantar flexion (ankle); frontal plane represents the eversion/inversion (ankle); horizontal plane represents the internal/external (knee). * indicates the significance level $p < 0.05$. Effect size (Cohen's d), small: $0.2 \leq d < 0.5$, middle: $0.5 \leq d < 0.8$ and large: $d > 0.8$.

4. Discussion

The purpose of this study was to investigate the effect of fatigue on postural control and biomechanical characteristic of lunges. The results demonstrated that: (i) the postural control was impaired within 10 min after fatigue, (ii) special postural variables and plantar sub-regions were more sensitive to fatigue, and (iii) the changes of kinematics and kinetics of lunges were consistent with postural impairment.

Contrary to studies that induced fatigue using repetitive isokinetic or isometric contractions, the current study used repeated forward lunging as the fatigue protocol. This way, the musculoskeletal system, load and angular velocity of the lower extremity joints were consistent with a badminton game. To a certain degree, this can better represent the state of fatigue in daily training and matches. Results showed that HR (HR/HRmax: 87.8~99.6%, mean $\pm$ SD: 93.5 $\pm$ 3.3%), BL (8.8~18.2 mmol/L, mean $\pm$ SD: 13.8 $\pm$ 2.7 mmol/L) and RPE (14~20, almost all data were greater than or equal to 18, except for one that was 14 and another that was 17; mean $\pm$ SD: 18.1 $\pm$ 1.5) increased immediately after the fatigue protocol. In combination with the results of the statistical analysis, it is reasonable to consider that the fatigue protocol induced fatigue in these participants.

Considering the effect of fatigue recovery, we measured the HR, BL and RPE at the 6th and 9th minutes after the fatigue protocol and performed the tests within 10 min. Despite HR, BL and RPE decreasing and being less than the criteria at T6 and T9, the BL was greater than 8 mmol/L at T9. Additionally, results of the statistical analysis also showed a significant increase at T6 and T9 for HR, BL and RPE. Consequently, we considered that the participants were fatigue during the test processing.

The postural control was impaired after the fatigue protocol. It is consistent with the conclusion that fatigue would minimize the ability to keep balance [3–5] and increase postural sway. However, a study [22] that adopted the Bosco protocol found that muscle fatigue of the lower limbs would affect vertical jump (VJ) performance, with no effect on balance. The discrepancy with the current study may indicate that the method to test the postural control differs. While VJ testing is the most common tool to explore lower-body power and strength in all sports [22], it can not follow the different movement strategies with similar total power output. Another study [19] also took repeating forward lunges as the fatigue protocol and showed changes in movement control and strategy during the lunge tasks, with no significant differences between pre- and post-fatigue for the Y-balance test. Such inconformity may be caused by the sports level of the participants [7] and the insensitivity of those clinical scores to the changes of performance.

It is worth noting that the effects of fatigue were not in the postural variables of the entire region, but those of plantar sub-regions. The rearfoot and forefoot were the sensitive regions for EO (area: $p = 0.05$; distance in ML: $p < 0.001$; distance in AP: $p = 0.029$) and EC (area: $p = 0.04$; displacement: $p = 0.006$; distance in ML: $p < 0.001$; distance in AP: $p = 0.035$), respectively. The medial of the forefoot is the most sensitive region to the fatigue protocol, with significant increase in displacement (EO: $p = 0.046$; EC: $p = 0.035$), distance in ML (EO: $p = 0.001$; EC: $p = 0.002$) and AP (EO: $p = 0.031$; EC: $p = 0.004$). As an asymmetrical movement, the dominant leg needs to complete the landing to the ground with the heel, lunging, supporting, braking and taking off. There are different loads in different plantar sub-regions [15,23,24]. It is well known that plantar cutaneous receptors help us balance and stand upright. The sensory feedback of the big toe and forefoot play an important role in the balance control of single-leg standing [28]. However, without data, it cannot verify whether the changes of postural sway were affected by the plantar cutaneous receptors. A study has confirmed that adding body load modified the vibratory sensation of the foot's sole, and it also significantly increased the CoP surface and lateral deviation [29]. In addition, the decreased ability of the musculoskeletal system caused by fatigue may be another reason for an increase of postural sway. In this study, peak powers of the knee (flexion and extension) and hip (extension) all decreased significantly. Similar results were found in a previous study [30]. Consequently, it may indicate that fatigue is only one of the factors contributing to the posture disorder.

Furthermore, among these variables, distance in ML is the most sensitive one to fatigue. During lunging, the gastrocnemius is the major one. A study has confirmed that the fatigue of gastrocnemius muscle had more obvious influence on the balance stability of the ML directions [31]. However, another study reported that the impairments in postural control were more evident in the AP direction when the plantar flexor and dorsiflexor muscles were affected [32]. Further work should focus on the function of these muscles.

Contrary to the negative effect of fatigue [14], proper neuromuscular training can improve the awareness of lower-limb joint and posture control [33]. Thus, studies of the influence of fatigue on the postural ability may provide a reference for reasonable and scientific training.

No differences were found for peak impact for the initial, secondary and drive-off phases. However, the timing of these variables was earlier than pre-fatigue, which can be explained by the decrease in postural control. After fatigue, the duration of III, between peak knee flexion and peak drive-off impact (PF2-PAK), increased significantly (pre: 27.3 ± 10.7 T%, post: 43.4 ± 15.2 T%, $p = 0.002$). This means that players spent more time preparing for drive-off. During this sub-phase, larger knee RoM, combined with the smaller knee joint moment in the sagittal plane, stands for the decrease of muscle strength. In addition, more attention should be paid to the angle of the ankle in frontal region. Smaller ankle eversion during stance might be a potential contributor to the injury risk of the ankle [8,15]. Except for the decrease in ankle eversion after fatigue, smaller joint moment of ankle in frontal ($p = 0.033$) is found, which means the induced ability of evertors around ankle. This may be used to explain why CoP sway in ML is more sensitive to fatigue. In badminton, there are higher rates of injury in the ankle and knee. Overuse injury caused by high-intensity training and repetitive movement is the main type of injury [8,9]. Lunging is an important footwork, with high-intensity use during training and competition [17]. During stance phase of the lunge, smaller knee joint moment in the sagittal ($p < 0.05$) may link to induced extensor muscle strength around the knee after fatigue. In addition, it is worth considering the changes in the hip. Before heel contact, sufficient flexion in the hip is important for a lunge. However, the smaller flexion angle of the hip at initial contact ($p = 0.044$) and hip joint moment in the sagittal plane are reported after fatigue. Furthermore, during drive-off, less power in the knee and hip in the sagittal plane may illustrate the decrease in control of the postural region. These suggest that specific muscles around the lower leg joints should be improved to maintain lunge performance, especially post-fatigue.

Considering the findings of this study, it did have a few limitations that should be considered when interpreting the results. Firstly, in the present study, only the dominant leg was tested for the single-leg balance test. Fatigue responses of non-dominant leg [34,35] and dynamic balance should be considered in further studies. Secondly, five seconds may give limited information, although a five-second delay was set for the postural test and one primary investigator performed all the tests. Further work should be conducted so as to determine the duration of the influence of fatigue recovery on postural control. Finally, all tests were performed in a simulated badminton court, and the participants were only college-aged male badminton players, with a limited sample size. Players of different genders, ages and sports levels may show different performance skills in badminton and may have different fatigue responses. A future study should take consideration of these differences.

5. Conclusions

In summary, the fatigue protocol had a significantly negative effect on postural control and biomechanical characteristic of lunge. The changes in biomechanical characteristic were consistent with the impairment of postural control. CoP sway of plantar sub-regions assessed by pressure plate system showed a significant response to fatigue. It provided us with the possibility of using postural sway data to monitor the state of motion, and to make an appropriate and scientific training plan to improve control ability and reduce the incidence of injury. On the other hand, future studies should extend the postural test time, combined with the EMG and dynamic balance test for a better understanding. Notice that avoiding the effect of fatigue recovery is important.

Author Contributions: Y.D. designed, carried out the experiments, analyzed the data and wrote the paper. Y.F. designed the experiments, and reviewed and revised the paper. All authors have read and agreed to the published version of the manuscript.

Funding: This research was funded by the National Natural Science Foundation of China (NSFC), grant No.: T2288101, U20A20390 and 11827803.

Institutional Review Board Statement: The study was conducted in accordance with the Declaration of Helsinki, and approved by the Ethics Committee of Beihang University (No. BM201900077).

Informed Consent Statement: Informed consent was obtained from all subjects involved in the study.

Data Availability Statement: Not applicable.

Acknowledgments: We thank all the participants in this study. Thank the NSFC.

Conflicts of Interest: The authors declare no conflict of interest.

References

1. Gandevia, S.C. Spinal and Supraspinal Factors in Human Muscle Fatigue. *Physiol. Rev.* **2001**, *81*, 1725–1789. [CrossRef] [PubMed]
2. Miura, K.; Ishibashi, Y.; Tsuda, E.; Okamura, Y.; Otsuka, H.; Toh, S. The Effect of Local and General Fatigue on Knee Proprioception. *Arthroscopy* **2004**, *20*, 414–418. [CrossRef] [PubMed]
3. Greig, M.; Walker-Johnson, C. The influence of soccer-specific fatigue on functional stability. *Phys. Ther. Sport* **2007**, *8*, 185–190. [CrossRef]
4. Brito, J.; Fontes, I.; Ribeiro, F.; Raposo, A.; Krustrup, P.; Rebelo, A. Postural stability decreases in elite young soccer players after a competitive soccer match. *Phys. Ther. Sport* **2012**, *13*, 175–179. [CrossRef]
5. Bisson, E.J.; Remaud, A.; Boyas, S.; Lajoie, Y.; Bilodeau, M. Effects of fatiguing isometric and isokinetic ankle exercises on postural control while standing on firm and compliant surfaces. *J. NeuroEng. Rehabil.* **2012**, *9*, 39. [CrossRef]
6. Arndt, A.; Ekenman, I.; Westblad, P.; Lundberg, A. Effects of fatigue and load variation on metatarsal deformation measured in vivo during barefoot walking. *J. Biomech.* **2002**, *35*, 621–628. [CrossRef]
7. Engebretsen, L.; Soligard, T.; Steffen, K.; Alonso, J.M.; Aubry, M.; Budgett, R.; Dvorak, J.; Jegathesan, M.; Meeuwisse, W.H.; Mountjoy, M.; et al. Sports injuries and illnesses during the London Summer Olympic Games 2012. *Br. J. Sports Med.* **2013**, *47*, 407–414. [CrossRef]
8. Shariff, A.H.; George, J.; Ramlan, A.A. Musculoskeletal injuries among Malaysian badminton players. *Singap. Med. J.* **2009**, *50*, 21–23.
9. Bickley, C.; Linton, J.; Sullivan, E.; Mitchell, K.; Slota, G.; Barnes, D. Comparison of simultaneous static standing balance data on a pressure mat and force plate in typical children and in children with cerebral palsy. *Gait Posture* **2019**, *67*, 91–98. [CrossRef]
10. Murray, L.; Beaven, C.M.; Hébert-Losier, K. The effects of running a 12-km race on neuromuscular performance measures in recreationally competitive runners. *Gait Posture* **2019**, *70*, 341–346. [CrossRef]
11. Behan, F.P.; Willis, S.; Pain, M.T.G.; Folland, J.P. Effects of football simulated fatigue on neuromuscular function and whole-body response to disturbances in balance. *Scand. J. Med. Sci. Sports* **2018**, *28*, 2547–2557. [CrossRef]
12. Barber-Westin, S.D.; Noyes, F.R. Effect of Fatigue Protocols on Lower Limb Neuromuscular Function and Implications for Anterior Cruciate Ligament Injury Prevention Training: A Systematic Review. *Am. J. Sports Med.* **2017**, *45*, 3388–3396. [CrossRef]
13. Paillard, T. Effects of general and local fatigue on postural control: A review. *Neurosci. Biobehav. Rev.* **2012**, *36*, 162–176. [CrossRef]
14. Roth, R.; Donath, L.; Zahner, L.; Faude, O. Acute Leg and Trunk Muscle Fatigue Differentially Affect Strength, Sprint, Agility, and Balance in Young Adults. *J. Strength Cond. Res.* **2019**, *35*, 2158–2164. [CrossRef]
15. Mei, Q.; Gu, Y.; Fu, F.; Fernandez, J. A biomechanical investigation of right-forward lunging step among badminton players. *J. Sports Sci.* **2017**, *35*, 457–462. [CrossRef]
16. Luiz, B.; Bedo, S.; Rodrigues, D.; Moraes, R.; Kalva-filho, C.A.; Will-de-lemos, T.; Roberto, P.; Santiago, P. The rapid recovery of vertical force propulsion production and postural sway after a specific fatigue protocol in female handball athletes. *Gait Posture* **2020**, *77*, 52–58. [CrossRef]
17. Huang, M.-T.; Lee, H.-H.; Lin, C.-F.; Tsai, Y.-J.; Liao, J.-C. How does knee pain affect trunk and knee motion during badminton forehand lunges? *J. Sports Sci.* **2014**, *32*, 690–700. [CrossRef]
18. Pincivero, D.M.; Aldworth, C.; Dickerson, T.; Petry, C.; Shultz, T. Quadriceps-hamstring EMG activity during functional, closed kinetic chain exercise to fatigue. *Eur. J. Appl. Physiol.* **2000**, *81*, 504–509. [CrossRef]
19. Valldecabres, R.; de Benito, A.M.; Littler, G.; Richards, J. An exploration of the effect of proprioceptive knee bracing on biomechanics during a badminton lunge to the net, and the implications to injury mechanisms. *PeerJ* **2018**, *6*, e6033. [CrossRef]
20. Kuntze, G.; Sellers, W.I.; Mansfield, N.J. Bilateral ground reaction forces and joint moments for lateral sidestepping and crossover stepping tasks. *J. Sports Sci. Med.* **2009**, *8*, 1–8.
21. Pau, M.; Ibba, G.; Attene, G. Fatigue-induced balance impairment in young soccer players. *J. Athl. Train.* **2014**, *49*, 454–461. [CrossRef] [PubMed]

22. Cooper, C.N.; Dabbs, N.C.; Davis, J.; Sauls, N.M. Effects of Lower-Body Muscular Fatigue on Vertical Jump and Balance Performance. *J. Strength Cond. Res.* **2018**, *34*, 2903–2910. [CrossRef] [PubMed]

23. Hong, Y.; Wang, S.J.; Lam, W.K.; Cheung, J.T.M. Kinetics of badminton lunges in four directions. *J. Appl. Biomech.* **2014**, *30*, 113–118. [CrossRef] [PubMed]

24. Fu, W.J.; Liu, Y.; Wei, Y. The characteristics of plantar pressure in typical footwork of badminton. *Footwear Sci.* **2009**, *1*, 113–115. [CrossRef]

25. Du, Y.; Fan, Y. Changes in the Kinematic and Kinetic Characteristics of Lunge Footwork during the Fatiguing Process. *Appl. Sci.* **2020**, *10*, 8703. [CrossRef]

26. Borg, G.A.V. Psychophysical bases of perceived exertion. *Med. Sci. Sports Exerc.* **1982**, *14*, 377–381. [CrossRef]

27. Cappozzo, A.; Catani, F.; Croce, U.D.; Leardini, A. Position and orientation in space of bones during movement: Anatomical frame definition and determination. *Clin. Biomech.* **1995**, *10*, 171–178. [CrossRef]

28. Meyer, P.F.; Oddsson, L.I.E.; De Luca, C.J. The role of plantar cutaneous sensation in unperturbed stance. *Exp. Brain Res.* **2004**, *156*, 505–512. [CrossRef]

29. Jammes, Y.; Ferrand, E.; Fraud, C.; Boussuges, A.; Weber, J.P. Adding body load modifies the vibratory sensation of the foot sole and affects the postural control. *Mil. Med. Res.* **2018**, *5*, 28. [CrossRef]

30. James, C.R.; Scheuermann, B.W.; Smith, M.P. Effects of two neuromuscular fatigue protocols on landing performance. *J. Electromyogr. Kinesiol.* **2010**, *20*, 667–675. [CrossRef]

31. Nam, H.S.; Park, D.S.; Kim, D.H.; Kang, H.J.; Lee, D.H.; Lee, S.H.; Her, J.G.; Woo, J.H.; Choi, S.Y. The relationship between muscle fatigue and balance in the elderly. *Ann. Rehabil. Med.* **2013**, *37*, 389–395. [CrossRef] [PubMed]

32. Gribble, P.A.; Hertel, J. Effect of lower-extremity muscle fatigue on postural control. *Arch. Phys. Med. Rehabil.* **2004**, *85*, 589–592. [CrossRef] [PubMed]

33. Vitale, J.A.; La Torre, A.; Banfi, G.; Bonato, M. Effects of an 8-weeks body-weight neuromuscular training on dynamic-balance and vertical jump performances in elite junior skiing athletes: A randomized controlled trial. *J. Strength Cond. Res.* **2018**, *32*, 911–920. [CrossRef] [PubMed]

34. Masu, Y.; Muramatsu, K.; Hayashi, N. Characteristics of sway in the center of gravity of badminton players. *J. Phys. Ther. Sci.* **2014**, *26*, 1671–1674. [CrossRef]

35. Ghram, A.; Young, J.D.; Soori, R.; Behm, D.G. Unilateral Knee and Ankle Joint Fatigue Induce Similar Impairment to Bipedal Balance in Judo Athletes. *J. Hum. Kinet.* **2019**, *66*, 7–18. [CrossRef]

Article

Effect of Different Landing Heights and Loads on Ankle Inversion Proprioception during Landing in Individuals with and without Chronic Ankle Instability

Ming Kang [1], Tongzhou Zhang [1], Ruoni Yu [2], Charlotte Ganderton [3], Roger Adams [4] and Jia Han [5,*]

1. School of Exercise and Health, Shanghai University of Sport, Shanghai 200438, China
2. School of Medicine, Jinhua Polytechnic, Jinhua 321000, China
3. Faculty of Health, Arts and Design, Swinburne University of Technology, Hawthorn, VIC 3122, Australia
4. Research Institute for Sport and Exercise, University of Canberra, Canberra, ACT 2234, Australia
5. College of Rehabilitation Sciences, Shanghai University of Medicine and Health Sciences, Shanghai 201318, China
* Correspondence: jia.han@canberra.edu.au

Abstract: Proprioception is essential for neuromuscular control in relation to sport injury and performance. The effect of landing heights and loads on ankle inversion proprioceptive performance in individuals with or without chronic ankle instability (CAI) may be important but are still unclear. Forty-three participants (21 CAI and 22 non-CAI) volunteered for this study. The Ankle Inversion Discrimination Apparatus for Landing (AIDAL), with one foot landing on a horizontal surface and the test foot landing on an angled surface ($10°$, $12°$, $14°$, $16°$), was utilized to assess ankle proprioception during landing. All participants performed the task from a landing height of 10 cm and 20 cm with 100% and 110% body weight loading. The four testing conditions were randomized. A repeated measures ANOVA was used for data analysis. The result showed that individuals with CAI performed significantly worse across the four testing conditions ($p = 0.018$). In addition, an increased landing height ($p = 0.010$), not loading ($p > 0.05$), significantly impaired ankle inversion discrimination sensitivity. In conclusion, compared to non-CAI, individuals with CAI showed significantly worse ankle inversion proprioceptive performance during landing. An increased landing height, not loading, resulted in decreased ankle proprioceptive sensitivity. These findings suggest that landing from a higher platform may increase the uncertainty of judging ankle positions in space, which may increase the risk of ankle injury.

Keywords: proprioception; ankle sprain; chronic ankle instability; landing

Citation: Kang, M.; Zhang, T.; Yu, R.; Ganderton, C.; Adams, R.; Han, J. Effect of Different Landing Heights and Loads on Ankle Inversion Proprioception during Landing in Individuals with and without Chronic Ankle Instability. *Bioengineering* **2022**, *9*, 743. https://doi.org/10.3390/bioengineering9120743

Academic Editors: Rui Zhang and Wei-Hsun Tai

Received: 14 October 2022
Accepted: 16 November 2022
Published: 30 November 2022

Publisher's Note: MDPI stays neutral with regard to jurisdictional claims in published maps and institutional affiliations.

1. Introduction

Ankle sprain is a common sports injury [1] that usually occurs during jump and landing activities [2–5] in a foot-inversion position [6,7] and may lead to chronic ankle instability (CAI) [8]. Proprioception is fundamental for neuromuscular control in ankle injury, and can be defined as the ability of an individual to integrate sensory signals from mechanoreceptors to perceive the location and spatial movement of body parts [9]. Studies have shown that neuromuscular control is deficient and ankle inversion proprioception to be significantly impaired and that the lower limb proximal muscle activity pattern is altered in individuals with CAI [10,11], especially during a landing task [12]. In some sports, such as cross-country running [13], participants may carry weights and land from different heights. It is unknown to what extent these factors may affect ankle proprioception during jump landing. Thus, exploring ankle inversion proprioception during landing is essential to understand the sensorimotor mechanisms underlying ankle sprains, and may inform prevention and rehabilitation of ankle sprain.

Previous biomechanical studies [14,15] suggests that significant biomechanical changes in movement patterns, muscle activation, and muscle mechanics occur in lower extremities when landing at different landing heights. Wang [16] reported that as landing height increases, angular displacement of the ankle, knee, and hip joints also increases, and lower extremity injuries are more likely to occur during landing. In addition, higher landing heights may lead to an increase in the velocity of the foot on landing, which is highly susceptible to injury of the foot and ankle complex [17], especially given that individuals with CAI have a delayed response of valgus muscle [18]. Fong [19] reported that within 0.11 s after the foot strike the ankle is in a position where it could be injured. Although these empirical studies indicate that motor behavior alters at different landing heights, it remains unclear whether the perceptual systems will also change. Although most current proprioceptive testing methods lack the ecological validity to carry out proprioceptive assessment during landing [9], Han developed the Ankle Inversion Discrimination Apparatus for Landing (AIDAL) [12] to make this measurement task achievable. We speculate whether the height of the landing affects the proprioceptive system, which further affects ankle stability.

Load is known to affect perceptual and motor system performance by altering postural control and proprioception. For instance, extra body weight can reduce postural-stability control and the sensory pathways from the foot sole [20,21]. However, weight-bearing can improve proprioceptive performance. The proprioceptive sensitivity of the knee joint was significantly better in full weight-bearing conditions than partial weight-bearing of the lower limb [22]. Considering the crucial significance of proprioception for postural control [23,24], we wondered whether the excess body load would have the effect on ankle proprioception during landing.

CAI is a condition characterized by pain, weakness, reduced ankle range of motion, perceived ankle "giving way" sensation, and proprioception deficit, which may lead to recurrent ankle sprains [25–28]. Although the soft tissues of the foot (e.g., fascia, muscles) perform a crucial role in maintaining the ankle stability, recurrent ankle sprains disrupt the original structure. Studies have shown that the thickness of the plantar fascia is reduced after lateral ankle sprain [29]. Moreover, individuals with CAI demonstrate sensorimotor insufficiencies and proprioception deficits [10,30], as well as biomechanical variations in lower limb movement patterns, motor strategies, muscle activation, and leg stiffness control during landing [24,31–34]. However, it is unclear whether a change in landing height and loading may have different effects on individuals with and without CAI. Evidence has suggested that the mechanoreceptors around the ankle joint are likely to be damaged in CAI [35], and so patients had diminished in neuromuscular control, especially ankle proprioception [5,10,12]. In addition, recent neuroimaging studies have found that this specific population also shows central change when performing proprioceptive balance control task [36–38]. These findings suggest that individuals with CAI may perform differently compared to their non-CAI counterparts.

Accordingly, the aim of this study was to test the ankle inversion proprioception of individuals with and without CAI when they landed at different heights with different loads. We hypothesized that an increased landing height may reduce ankle proprioception, but extra loads may reverse this, and individuals with CAI would perform significantly worse than those without CAI.

2. Materials and Methods

2.1. Study Design

A cross-sectional study was performed between June and July 2021 according to the Strengthening the Reporting of Observational Studies in Epidemiology (STROBE) recommendations.

2.2. Participants

Forty-three participants were recruited (21 CAI and 22 non-CAI). To be eligible, the CAI participants must have had the following: (i) at least one ankle sprain that caused an inflammatory reaction (pain, swelling, etc.) in the previous 12 months; (ii) at least two

episodes of ankle instability "giving way" or repeated sprains within 6 months before the test; (iii) a Cumberland Ankle Instability Tool score (CAIT) [39] of <24; and (iv) not have had an ankle injury within 3-months of being tested. These inclusion criteria align with the recommendations of the International Ankle Consortium [39]. The non-CAI participants had no subjective reports of ankle instability, recurrent ankle sprains, and neurological or motor dysfunction. All participants were excluded from the study if they had a history of lower limb surgery, fracture, or any acute injury of other joints of the lower limb in the 3-months prior to the commencement of the study. The latest ankle sprain was 15 weeks on average before the test for the CAI group.

The study was conducted in accordance with the Declaration of Helsinki and was approved by the Shanghai University of Sport Ethics Committee (102772021RT073). Written informed consent was obtained from participants before data collection.

2.3. Apparatus

This study utilized the AIDAL to measure the acuity of ankle inversion proprioception during landing [12]. This method has shown good test–retest reliability and validity in distinguishing individuals with and without CAI [40]. The AIDAL (Figure 1) consists of three parts: the take-off platform (A/D), the horizontal landing platform for the supporting foot (B), and the tilted landing platform for the testing foot (C). The four different angles of ankle inversion were generated by 4 wedged landing platforms: inversion 1 = 10°, inversion 2 = 12°, inversion 3 = 14°, and inversion 4 = 16°. The landing heights were 10 cm (A to B) and 20 cm (D to B) (Figures 1 and 2, a and b).

Before the AIDAL's data collection, each participant had three rounds of familiarization of the 4 possible ankle inversions in order (12 trials in total), and they were required to remember the four different ankle inversions during the familiarization session. Participants then undertook 40 trials of testing, with 10 for each inversion position presented randomly. Participants were required to make an absolute judgment about the ankle inversion angle on each testing trial, and the numbers (either 1, 2, 3, or 4) with which the participant responded were collected, and without feedback being given as to the correctness of their judgements.

The weight-adjustable vest (Figure 2c,d) was used to provide 10% of extra body weight during the AIDAL proprioception test [22].

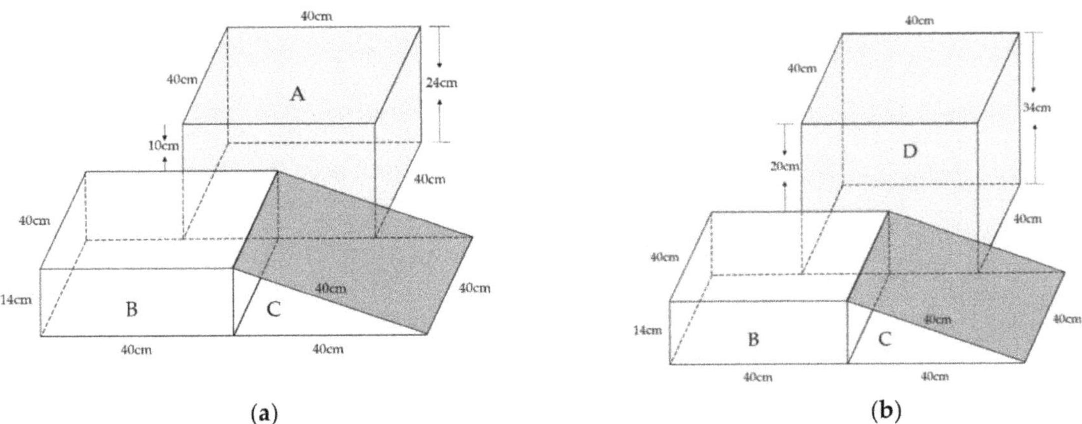

Figure 1. The two different heights of AIDAL: (**a**) 10 cm heights between A and B; (**b**) 20 cm heights between D and B.

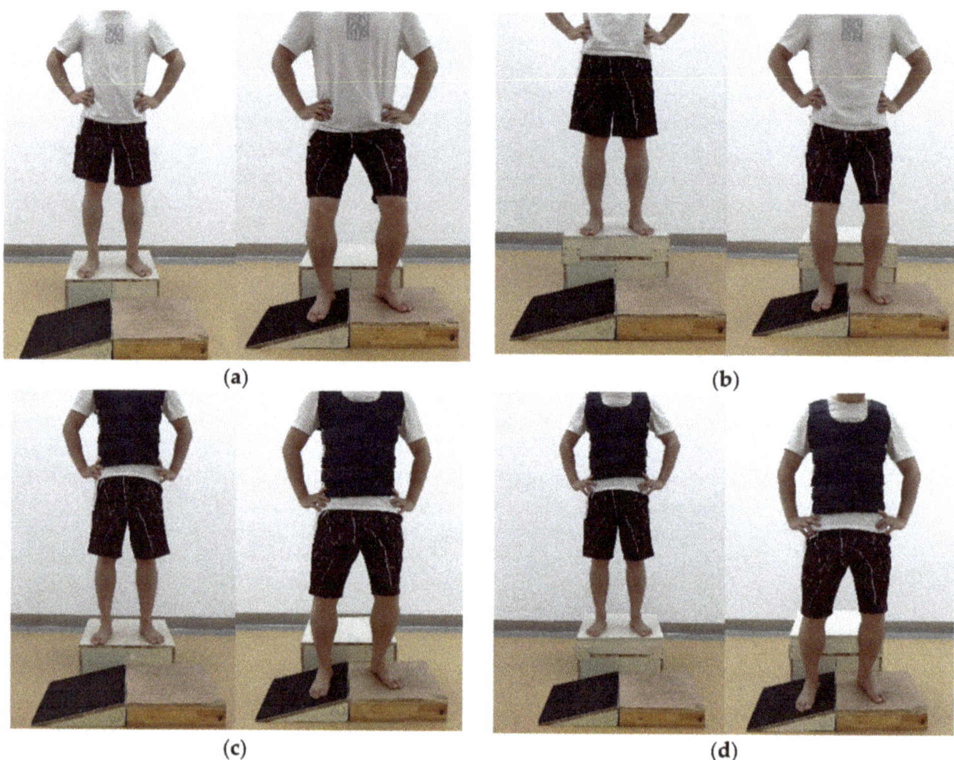

Figure 2. The different landing conditions. (**a**) load 100% and 10 cm heights; (**b**) load 100% and 20 cm heights; (**c**) load 110% and 10 cm heights; (**d**) load 110% and 20 cm heights.

2.4. Procedures

The experiment was conducted in a university laboratory. All participants were tested with bare feet. For the non-CAI group, the testing foot was randomly selected, and for the CAI group, with unilateral CAI (n = 10), we tested the affected ankle, and with bilateral CAI (n = 11), we tested the ankle with a lower CAIT score. The flow chart of this study is shown below (Figure 3). All participants were tested from a jumping height of 10 cm and 20 cm without extra load (100% body weight), and with an extra 10% of body weight (110% body weight) (Figure 2). The four testing conditions were randomized with a 15-min break between testing sessions. During the test, participants were instructed to keep their head and eyes forward to eliminate visual information about the landing platforms. A single examiner who was blinded to participants' ankle stability status performed all experiments.

2.5. Data Analysis

SPSS version 25 (Armonk, NY, USA) was used for data analysis and a *p* value of 0.05 or less was used to determine statistical significance. A total of 40 presentations of ankle inversion and related participant responses were inputted into SPSS to generate a receiver operating curve (ROC), and the area under the curve (AUC) was calculated as the ankle proprioception discrimination sensitivity. The AUC value ranges between 0.5 and 1. A higher value represents more accurate proprioception sensitivity.

Given that the data for the two groups were normally distributed, to determine the effect of CAI, landing heights and loads on ankle proprioception, a repeated measures ANOVA was conducted on the AUC scores.

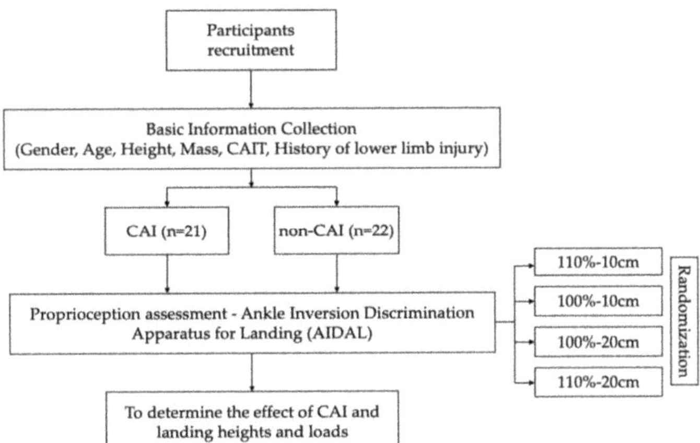

Figure 3. The flow chart of this study. A total of 100% meant original body weight and 110% meant 10% extra body weight; 10 cm and 20 cm indicated landing heights.

3. Results

Demographic information of the included participants is reported in Table 1. The repeated measures ANOVA showed CAI and the landing height main effects. Specifically, the CAI group performed significantly worse than the non-CAI group (F = 6.120, p = 0.018, partial η^2 = 0.130) and the ankle proprioception AUC scores of 20 cm landing height were significantly lower than that of 10 cm landing height (F = 7.216, p = 0.010, partial η^2 = 0.150) (Table 2 and Figure 4). However, there was not a significant load main effect (F < 0.001, p = 0.995, partial η^2 < 0.001) or an interaction effect of height, load, and presence of CAI (Table 2).

Table 1. Participant demographic information (Mean ± SD).

Characteristic	Group		Difference between Groups
	CAI	**Non-CAI**	
N	21	22	-
Gender	M10 F11	M11 F11	-
Age (y)	23.4 ± 3.2	24.1 ± 2.1	$t = -0.804, p = 0.426$
Height (cm)	171.3 ± 8.2	169.1 ± 6.3	$t = 0.972, p = 0.337$
Mass (kg)	65.6 ± 11.6	64.6 ± 9.0	$t = 0.268, p = 0.790$
CAIT score	15.6 ± 4.9	28.6 ± 1.8	$t = -11.606, p = 0.000$

SD = standard deviation, CAI = chronic ankle instability, N = Number, M/F = male/female, CAIT = Cumberland Ankle Instability Tool.

Table 2. A repeated measures ANOVA of the average AUC scores for landing height and group.

	F	**p**	**Partial η^2**
10 cm vs. 20 cm	7.216	0.010 *	0.150
CAI vs. Non-CAI	6.120	0.018 *	0.130
100% vs. 110%	<0.001	0.995	<0.001
Height × Load	1.874	0.178	0.044
Height × CAI	<0.001	0.984	<0.001
Load × CAI	2.244	0.142	0.052
Height × Load × CAI	0.001	0.972	<0.001

× = interaction between loads and heights. * = p < 0.05 between groups.

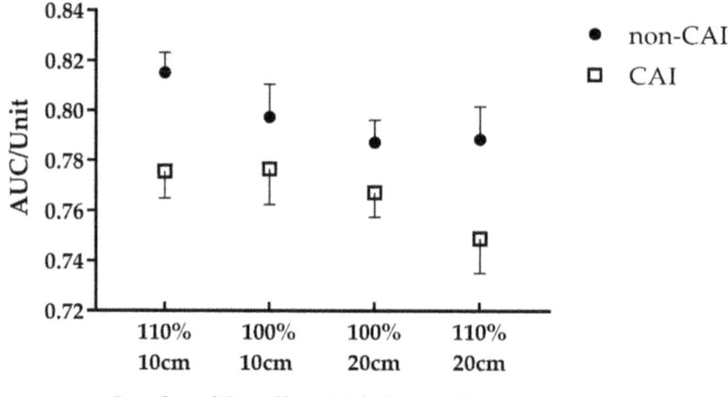

Load and Landing Height conditions

Figure 4. Differences in ankle inversion discrimination tested during landing heights and loads between individuals with and without CAI. A total of 100% indicated original body weight and 110% indicated 10% extra body weight; 10 cm and 20 cm represented landing heights. Compared to non-CAI, CAI showed significantly worse ankle inversion proprioceptive performance during landing (F = 6.120, p = 0.018), and the AUC score decreased by increased landing height for all participants (F = 7.216, p = 0.010).

4. Discussion

Consistent with our hypothesis, we found that the overall ankle inversion proprioceptive performance of CAI patients was significantly worse than that of non-CAI participants, suggesting that CAI patients have impaired proprioceptive control when landing on an uneven surface. In addition, ankle proprioceptive discrimination sensitivity was significantly worsened by an increased landing height, but not loading (Figure 4), which was true for both CAI and non-CAI groups.

Our results showed that the proprioceptive acuity of CAI participants was significantly worse than those without CAI across the four different testing conditions. These findings are consistent with prior studies [12,41] and further supports the notion that CAI patients have impaired somatosensory control during landing on an inverted ankle [12]. Some studies have shown that individuals with CAI exhibit altered peak proximal muscle forces, force-generating capacities, as well as greater hip flexion and ankle inversion angles, and peak vertical ground reaction forces during landing tasks [42–44]. Our findings complement the substantial deviations in the lower limb motor output observed between CAI and non-CAI individuals and have shown that ankle proprioceptive input is also different between the two groups. According to the research of Waddington and Adams [45], even a 0.04° increase in inversion uncertainty has the potential to raise the probability of injury when landing on the inverted ankle from 1.2% to 1.22%. Although this 0.02% increase in injury seems low, it could become a significant influence in the occurrence of injury due to the fact that landings are numerous in sports activities. Therefore, the difference in proprioception between the CAI and non-CAI participants in this study has significant implications for ankle stability and may raise the risk of sprain. Given that ankle proprioception is fundamental for lower limb motor control [23], the difference in proprioceptive performance during landing found here may partially explain the motor output difference between CAI and non-CAI observed in previous studies [43,44]. Future research may explore if any rehabilitation program that targets ankle proprioception [46] could have positive effects on lower limb motor control in people with a post-ankle sprain that may or may not develop CAI.

In terms of the effect of landing height on lower limb biomechanics, research has shown that with increasing landing height, the ankle dorsiflexion, knee extension, and peak ankle plantarflexion moments are significantly altered in individuals with CAI [47]. The results of the current study show that an increased landing height could impair ankle inversion proprioception in both CAI and non-CAI participants. This evidence supports the notion that the larger the movement amplitude, the worse the proprioceptive performance [48], suggesting that landing from a higher place may be associated with increased noise in the perceptual systems of brain [48,49], so that ankle inversion movement discrimination sensitivity is decreased. Previous upper [48,50] and lower [51] limb proprioception studies have found that larger movement resulted in worse proprioceptive acuity. This is consistent with the findings of our result. A greater movement amplitude would generate more noise and increased uncertainty when judging limb positions in space. This finding suggests that higher jump–landing may increase the risk of ankle injury and partially explains why landing from a jump is one of the most common mechanisms for ankle injury [2], especially in ankle joint inversion landing conditions [52].

However, we found an extra 10% of body weight for the proprioceptive measurement did not differ from 100% body weight conditions. This evidence is contrary to the argument that ankle proprioception simply relied on mechanoreceptors around the foot–ankle complex [53]. Han et al. [9,54] argue that proprioception assessment methods can be classified into testing passively "imposed" and actively "obtained" proprioception. The imposed methods [53] believed that proprioception is completely reliant on information received from the peripheral proprioceptors mechanoreceptors (i.e., muscles, joints, and skin) [55]. In contrast, the actively obtained methods [56] were developed on the basis of an ecological validity concept that proprioceptive performance is not fully determined by passively-imposed proprioceptive signals from mechanoreceptors, but requires adaptive central processing of multiple sources of information [9]. If the "imposed" proprioception view is true, then both an increased landing height and load could further activate mechanoreceptors located around the foot–ankle complex, improving ankle proprioception. However, the results here did not support this notion. The possible reason is that an additional 10% of load may not have been large enough and that the central nervous system may have mechanisms to adapt to the noise generated by mild changes in weight-bearing conditions. In addition, although we required participants to land evenly, participants might not have achieved a balanced distribution of weight between the two feet, with more weight on the horizontal platform, thus this may have reduced the load on the inverted foot.

This study compared ankle inversion proprioception during landing in both CAI and non-CAI populations in the face of height and loading conditions. One of the limitations of this study was that we did not collect data about the physical activity and occupation of the participants included in this study, which may have an impact on the results. Furthermore, we did not quantify kinematic patterns so that precise changes in movement extent of the participants at different heights and loading states to be observed. We did not use any instruments to control whether participants had a balanced distribution of weight between the two feet, a feature which should be improved for future study. In addition, previous research has shown that female and male individuals with CAI performed differently on a range of functional tasks [57]. However, given the relatively small sample size of the current study, the sex differences in this proprioceptive task were not revealed. Furthermore, the participants involved in the current study were relatively young and it is unknown if the findings here can be generalized to other age groups.

5. Conclusions

Compared to non-CAI, individuals with CAI showed significantly worse ankle inversion proprioceptive performance during landing. An increased landing height, but not loading, resulted in decreased ankle proprioceptive sensitivity. These findings suggest that landing from a higher place may increase the uncertainty of judging ankle positions in space, and thus increases the risk of ankle injuries. Therefore, jump-landing exercise from different heights may be important for ankle injury prevention and rehabilitation. Future research may investigate the effects of a rehabilitation program targeting ankle inversion proprioception during landing and explore the peripheral and central mechanisms associated with these effects.

Author Contributions: Conceptualization, M.K. and J.H.; methodology, R.A.; software, M.K. and R.A.; validation, J.H. and R.A.; formal analysis, T.Z.; investigation, M.K. and T.Z.; resources, J.H.; data curation, M.K. and J.H.; writing—original draft preparation, M.K.; writing—review and editing, J.H., R.Y. and C.G.; visualization, C.G.; supervision, J.H.; project administration, M.K.; funding acquisition, J.H. All authors have read and agreed to the published version of the manuscript.

Funding: This research was funded by the National Natural Science Foundation of China, grant number 31870936; and the Program of Science and Technology Commission of Shanghai Municipality (Excellent Academic Leader (Youth) of Science and Technology Innovation Action Plan), grant number 20XD1423200.

Institutional Review Board Statement: The study was conducted in accordance with the Declaration of Helsinki and approved by the Ethics Committee of Shanghai University of Sport (protocol code: 102772021RT073 and date of approval: 24 May 2021).

Informed Consent Statement: Informed consent was obtained from all subjects involved in the study.

Data Availability Statement: The data presented in this study are available on request from the corresponding author. The data are not publicly available due to ethical or privacy restrictions.

Conflicts of Interest: The authors declare no conflict of interest.

References

1. Fong, D.T.; Hong, Y.; Chan, L.K.; Yung, P.S.; Chan, K.M. A systematic review on ankle injury and ankle sprain in sports. *Sports Med.* **2007**, *37*, 73–94. [CrossRef] [PubMed]
2. Lytle, J.B.; Parikh, K.B.; Tarakemeh, A.; Vopat, B.G.; Mulcahey, M.K. Epidemiology of Foot and Ankle Injuries in NCAA Jumping Athletes in the United States During 2009–2014. *Orthop. J. Sports Med.* **2021**, *9*, 2325967121998052. [CrossRef] [PubMed]
3. Waterman, B.R.; Owens, B.D.; Davey, S.; Zacchilli, M.A.; Belmont, P.J.J.J., Jr. The epidemiology of ankle sprains in the United States. *J. Bone Jt. Surg. Am.* **2010**, *92*, 2279–2284. [CrossRef]
4. McKay, G.D.; Goldie, P.A.; Payne, W.R.; Oakes, B.W. Ankle injuries in basketball: Injury rate and risk factors. *Br. J. Sports Med.* **2001**, *35*, 103–108. [CrossRef] [PubMed]
5. Lin, J.-Z.; Lin, Y.-A.; Tai, W.-H.; Chen, C.-Y. Influence of Landing in Neuromuscular Control and Ground Reaction Force with Ankle Instability: A Narrative Review. *Bioengineering* **2022**, *9*, 68. [CrossRef] [PubMed]
6. Panagiotakis, E.; Mok, K.M.; Fong, D.T.; Bull, A.M.J. Biomechanical analysis of ankle ligamentous sprain injury cases from televised basketball games: Understanding when, how and why ligament failure occurs. *J. Sci. Med. Sport* **2017**, *20*, 1057–1061. [CrossRef]
7. Fong, D.T.; Chan, Y.Y.; Mok, K.M.; Yung, P.S.; Chan, K.M. Understanding acute ankle ligamentous sprain injury in sports. *Sports Med. Arthrosc. Rehabil. Ther. Technol.* **2009**, *1*, 14. [CrossRef]
8. Doherty, C.; Bleakley, C.; Delahunt, E.; Holden, S. Treatment and prevention of acute and recurrent ankle sprain: An overview of systematic reviews with meta-analysis. *Br. J. Sports Med.* **2017**, *51*, 113–125. [CrossRef]
9. Han, J.; Waddington, G.; Adams, R.; Anson, J.; Liu, Y. Assessing proprioception: A critical review of methods. *J. Sport Health Sci.* **2016**, *5*, 80–90. [CrossRef]
10. Xue, X.; Ma, T.; Li, Q.; Song, Y.; Hua, Y. Chronic ankle instability is associated with proprioception deficits: A systematic review and meta-analysis. *J. Sport Health Sci.* **2021**, *10*, 182–191. [CrossRef]
11. Kazemi, K.; Arab, A.M.; Abdollahi, I.; López-López, D.; Calvo-Lobo, C. Electromyography comparison of distal and proximal lower limb muscle activity patterns during external perturbation in subjects with and without functional ankle instability. *Hum. Mov. Sci.* **2017**, *55*, 211–220. [CrossRef] [PubMed]
12. Han, J.; Yang, Z.; Adams, R.; Ganderton, C.; Witchalls, J.; Waddington, G. Ankle inversion proprioception measured during landing in individuals with and without chronic ankle instability. *J. Sci. Med. Sport* **2021**, *24*, 665–669. [CrossRef]
13. Hunt, K.J.; Hurwit, D.; Robell, K.; Gatewood, C.; Botser, I.B.; Matheson, G. Incidence and Epidemiology of Foot and Ankle Injuries in Elite Collegiate Athletes. *Am. J. Sports Med.* **2017**, *45*, 426–433. [CrossRef] [PubMed]
14. Nordin, A.D.; Dufek, J.S. Lower extremity variability changes with drop-landing height manipulations. *Res. Sports Med.* **2017**, *25*, 144–155. [CrossRef] [PubMed]
15. Wang, I.L.; Chen, Y.M.; Zhang, K.K.; Li, Y.G.; Su, Y.; Wu, C.; Ho, C.S. Influences of Different Drop Height Training on Lower Extremity Kinematics and Stiffness during Repetitive Drop Jump. *Appl. Bionics Biomech.* **2021**, *2021*, 5551199. [CrossRef] [PubMed]
16. Wang, I.L.; Wang, S.Y.; Wang, L.I. Sex differences in lower extremity stiffness and kinematics alterations during double-legged drop landings with changes in drop height. *Sports Biomech.* **2015**, *14*, 404–412. [CrossRef]

17. Gribble, P.A.; Bleakley, C.M.; Caulfield, B.M.; Docherty, C.L.; Fourchet, F.; Fong, D.T.; Hertel, J.; Hiller, C.E.; Kaminski, T.W.; McKeon, P.O.; et al. 2016 consensus statement of the International Ankle Consortium: Prevalence, impact and long-term consequences of lateral ankle sprains. *Br. J. Sports Med.* **2016**, *50*, 1493–1495. [CrossRef]
18. Hopkins, J.T.; Brown, T.N.; Christensen, L.; Palmieri-Smith, R.M. Deficits in peroneal latency and electromechanical delay in patients with functional ankle instability. *J. Orthop. Res.* **2009**, *27*, 1541–1546. [CrossRef]
19. Fong, D.T.; Hong, Y.; Shima, Y.; Krosshaug, T.; Yung, P.S.; Chan, K.M. Biomechanics of supination ankle sprain: A case report of an accidental injury event in the laboratory. *Am. J. Sports Med.* **2009**, *37*, 822–827. [CrossRef]
20. Wojciechowska-Maszkowska, B.; Borzucka, D. Characteristics of Standing Postural Control in Women under Additional Load. *Int. J. Environ. Res. Public Health* **2020**, *17*, 490. [CrossRef]
21. Jammes, Y.; Ferrand, E.; Fraud, C.; Boussuges, A.; Weber, J.P. Adding body load modifies the vibratory sensation of the foot sole and affects the postural control. *Mil. Med. Res.* **2018**, *5*, 28. [CrossRef] [PubMed]
22. Bullock-Saxton, J.E.; Wong, W.J.; Hogan, N. The influence of age on weight-bearing joint reposition sense of the knee. *Exp. Brain Res.* **2001**, *136*, 400–406. [CrossRef] [PubMed]
23. Han, J.; Anson, J.; Waddington, G.; Adams, R.; Liu, Y. The Role of Ankle Proprioception for Balance Control in relation to Sports Performance and Injury. *Biomed. Res. Int.* **2015**, *2015*, 842804. [CrossRef] [PubMed]
24. Simpson, J.D.; Stewart, E.M.; Macias, D.M.; Chander, H.; Knight, A.C. Individuals with chronic ankle instability exhibit dynamic postural stability deficits and altered unilateral landing biomechanics: A systematic review. *Phys. Ther. Sport* **2019**, *37*, 210–219. [CrossRef]
25. Vuurberg, G.; Hoorntje, A.; Wink, L.M.; van der Doelen, B.F.W.; van den Bekerom, M.P.; Dekker, R.; van Dijk, C.N.; Krips, R.; Loogman, M.C.M.; Ridderikhof, M.L.; et al. Diagnosis, treatment and prevention of ankle sprains: Update of an evidence-based clinical guideline. *Br. J. Sports Med.* **2018**, *52*, 956. [CrossRef]
26. Hertel, J.; Corbett, R.O. An Updated Model of Chronic Ankle Instability. *J. Athl. Train.* **2019**, *54*, 572–588. [CrossRef]
27. Morales, C.R.; Lobo, C.C.; Sanz, D.R.; Corbalán, I.S.; Ruiz, B.R.; López, D.L. The concurrent validity and reliability of the Leg Motion system for measuring ankle dorsiflexion range of motion in older adults. *PeerJ* **2017**, *5*, e2820. [CrossRef]
28. Cruz-Díaz, D.; Vega, R.L.; Osuna-Pérez, M.C.; Hita-Contreras, F.; Martínez-Amat, A. Effects of joint mobilization on chronic ankle instability: A randomized controlled trial. *Disabil. Rehabil.* **2015**, *37*, 601–610. [CrossRef]
29. Romero-Morales, C.; López-López, S.; Bravo-Aguilar, M.; Cerezo-Téllez, E.; Benito-de Pedro, M.; López, D.L.; Lobo, C.C. Ultrasonography Comparison of the Plantar Fascia and Tibialis Anterior in People with and Without Lateral Ankle Sprain: A Case-Control Study. *J. Manip. Physiol. Ther.* **2020**, *43*, 799–805. [CrossRef]
30. Delahunt, E.; Bleakley, C.M.; Bossard, D.S.; Caulfield, B.M.; Docherty, C.L.; Doherty, C.; Fourchet, F.; Fong, D.T.; Hertel, J.; Hiller, C.E.; et al. Clinical assessment of acute lateral ankle sprain injuries (ROAST): 2019 consensus statement and recommendations of the International Ankle Consortium. *Br. J. Sports Med.* **2018**, *52*, 1304–1310. [CrossRef]
31. Kaminski, T.W.; Knight, C.; Glutting, J.; Thomas, S.; Swanik, C.; Rosen, A. Differences in Lateral Drop Jumps from an Unknown Height Among Individuals With Functional Ankle Instability. *J. Athl. Train.* **2013**, *48*, 773–781. [CrossRef]
32. Lin, J.Z.; Lin, Y.A.; Lee, H.J. Are Landing Biomechanics Altered in Elite Athletes with Chronic Ankle Instability. *J. Sports Sci. Med.* **2019**, *18*, 653–662. [PubMed]
33. DeJong, A.F.; Koldenhoven, R.M.; Hertel, J. Hip biomechanical alterations during walking in chronic ankle instability patients: A cross-correlation analysis. *Sports Biomech.* **2022**, *21*, 460–471. [CrossRef] [PubMed]
34. Jeon, K.; Kim, K.; Kang, N. Leg stiffness control during drop landing movement in individuals with mechanical and functional ankle disabilities. *Sports Biomech.* **2022**, *21*, 1093–1106. [CrossRef] [PubMed]
35. Refshauge, K.M.; Kilbreath, S.L.; Raymond, J. The effect of recurrent ankle inversion sprain and taping on proprioception at the ankle. *Med. Sci. Sports Exerc.* **2000**, *32*, 10–15. [CrossRef]
36. Rosen, A.B.; Yentes, J.M.; McGrath, M.L.; Maerlender, A.C.; Myers, S.A.; Mukherjee, M. Alterations in cortical activation among individuals with chronic ankle instability during single-limb postural control. *J. Athl. Train.* **2019**, *54*, 718–726. [CrossRef] [PubMed]
37. Koren, Y.; Parmet, Y.; Bar-Haim, S. Treading on the unknown increases prefrontal activity: A pilot fNIRS study. *Gait Posture* **2019**, *69*, 96–100. [CrossRef]
38. Stuart, S.; Vitorio, R.; Morris, R.; Martini, D.N.; Fino, P.C.; Mancini, M. Cortical activity during walking and balance tasks in older adults and in people with Parkinson's disease: A structured review. *Maturitas* **2018**, *113*, 53–72. [CrossRef]
39. Gribble, P.A.; Delahunt, E.; Bleakley, C.; Caulfield, B.; Docherty, C.; Fourchet, F.; Fong, D.T.; Hertel, J.; Hiller, C.; Kaminski, T.; et al. Selection criteria for patients with chronic ankle instability in controlled research: A position statement of the International Ankle Consortium. *Br. J. Sports Med.* **2014**, *48*, 1014–1018. [CrossRef]
40. Hiller, C.E.; Refshauge, K.M.; Bundy, A.C.; Herbert, R.D.; Kilbreath, S.L. The Cumberland ankle instability tool: A report of validity and reliability testing. *Arch. Phys. Med. Rehabil.* **2006**, *87*, 1235–1241. [CrossRef]
41. Witchalls, J.B.; Waddington, G.; Adams, R.; Blanch, P. Chronic ankle instability affects learning rate during repeated proprioception testing. *Phys. Ther. Sport* **2014**, *15*, 106–111. [CrossRef] [PubMed]
42. Watabe, T.; Takabayashi, T.; Tokunaga, Y.; Watanabe, T.; Kubo, M. Copers exhibit altered ankle and trunk kinematics compared to the individuals with chronic ankle instability during single-leg landing. *Sports Biomech.* **2022**, 1–13. [CrossRef] [PubMed]

43. Yu, P.; Mei, Q.; Xiang, L.; Fernandez, J.; Gu, Y. Differences in the locomotion biomechanics and dynamic postural control between individuals with chronic ankle instability and copers: A systematic review. *Sports Biomech.* **2022**, *21*, 531–549. [CrossRef] [PubMed]

44. Kim, H.; Palmieri-Smith, R.; Kipp, K. Peak Forces and Force Generating Capacities of Lower Extremity Muscles During Dynamic Tasks in People With and Without Chronic Ankle Instability. *Sports Biomech.* **2022**, *21*, 487–500. [CrossRef] [PubMed]

45. Waddington, G.; Adams, R. Football boot insoles and sensitivity to extent of ankle inversion movement. *Br. J. Sports Med.* **2003**, *37*, 170–174, discussion 175. [CrossRef] [PubMed]

46. Han, J.; Luan, L.; Adams, R.; Witchalls, J.; Newman, P.; Tirosh, O.; Waddington, G. Can therapeutic exercises improve proprioception in chronic ankle instability? A systematic review and network meta-analysis. *Arch. Phys. Med. Rehabil.* **2022**, *103*, 2232–2244. [CrossRef]

47. Watanabe, K.; Koshino, Y.; Ishida, T.; Samukawa, M.; Tohyama, H. Energy dissipation during single-leg landing from three heights in individuals with and without chronic ankle instability. *Sports Biomech.* **2022**, *21*, 408–427. [CrossRef]

48. Goble, D.J. Proprioceptive acuity assessment via joint position matching: From basic science to general practice. *Phys. Ther.* **2010**, *90*, 1176–1184. [CrossRef]

49. Faisal, A.A.; Selen, L.P.; Wolpert, D.M. Noise in the nervous system. *Nat. Rev. Neurosci.* **2008**, *9*, 292–303. [CrossRef]

50. Han, J.; Adams, R.; Waddington, G.; Han, C.J.S.; Research, M. Proprioceptive accuracy after uni-joint and multi-joint patterns of arm-raising movements directed to overhead targets. *Somatosens. Mot. Res.* **2021**, *38*, 127–132. [CrossRef]

51. Symes, M.; Waddington, G.; Adams, R. Depth of ankle inversion and discrimination of foot positions. *Percept. Mot. Ski.* **2010**, *111*, 475–484. [CrossRef] [PubMed]

52. Fong, D.T.; Ha, S.C.; Mok, K.M.; Chan, C.W.; Chan, K.M. Kinematics analysis of ankle inversion ligamentous sprain injuries in sports: Five cases from televised tennis competitions. *Am. J. Sports Med.* **2012**, *40*, 2627–2632. [CrossRef] [PubMed]

53. Weerakkody, N.; Blouin, J.; Taylor, J.; Gandevia, S.C. Local subcutaneous and muscle pain impairs detection of passive movements at the human thumb. *J. Physiol.* **2008**, *586*, 3183–3193. [CrossRef]

54. Han, J.; Adams, R.; Waddington, G. "Imposed" and "obtained" ankle proprioception across the life span—Commentary on Djajadikarta et al. *J. Appl. Physiol.* **2020**, *129*, 533–534. [CrossRef] [PubMed]

55. Lephart, S.M.; Pincivero, D.M.; Giraldo, J.L.; Fu, F.H. The role of proprioception in the management and rehabilitation of athletic injuries. *Am. J. Sports Med.* **1997**, *25*, 130–137. [CrossRef] [PubMed]

56. Waddington, G.; Adams, R. Discrimination of active plantarflexion and inversion movements after ankle injury. *Aust. J. Physiother.* **1999**, *45*, 7–13. [CrossRef]

57. Lu, J.; Wu, Z.; Adams, R.; Han, J.; Cai, B. Sex differences in the relationship of hip strength and functional performance to chronic ankle instability scores. *J. Orthop. Surg. Res.* **2022**, *17*, 173. [CrossRef]

bioengineering

Article

Identification of Kinetic Abnormalities in Male Patients after Anterior Cruciate Ligament Deficiency Combined with Meniscal Injury: A Musculoskeletal Model Study of Lower Limbs during Jogging

Shuang Ren [1,†], Xiaode Liu [2,3,†], Haoran Li [2], Yufei Guo [3], Yuhan Zhang [3], Zixuan Liang [1], Si Zhang [1], Hongshi Huang [1], Xuhui Huang [3], Zhe Ma [3], Qiguo Rong [2,*] and Yingfang Ao [1,*]

1   Beijing Key Laboratory of Sports Injuries, Department of Sports Medicine, Institute of Sports Medicine of Peking University, Peking University Third Hospital, Beijing 100191, China
2   Department of Mechanics and Engineering Science, College of Engineering, Peking University, Beijing 100871, China
3   Intelligent Science & Technology Academy of CASIC, Beijing 100043, China
*   Correspondence: qrong@pku.edu.cn (Q.R.); aoyingfang@163.com (Y.A.)
†   These authors contributed equally to this work.

Citation: Ren, S.; Liu, X.; Li, H.; Guo, Y.; Zhang, Y.; Liang, Z.; Zhang, S.; Huang, H.; Huang, X.; Ma, Z.; et al. Identification of Kinetic Abnormalities in Male Patients after Anterior Cruciate Ligament Deficiency Combined with Meniscal Injury: A Musculoskeletal Model Study of Lower Limbs during Jogging. *Bioengineering* **2022**, *9*, 716. https://doi.org/10.3390/bioengineering9110716

Academic Editors: Rui Zhang, Wei-Hsun Tai and Stuart Goodman

Received: 15 October 2022
Accepted: 17 November 2022
Published: 19 November 2022

Publisher's Note: MDPI stays neutral with regard to jurisdictional claims in published maps and institutional affiliations.

Abstract: There is little known about kinetic changes in anterior cruciate ligament deficiency (ACLD) combined with meniscal tears during jogging. Therefore, 29 male patients with injured ACLs and 15 healthy male volunteers were recruited for this study to investigate kinetic abnormalities in male patients after ACL deficiency combined with a meniscal injury during jogging. Based on experimental data measured by an optical tracking system, a subject-specific musculoskeletal model was employed to estimate the tibiofemoral joint kinetics during jogging. Between-limb and interpatient differences were compared by the analysis of variance. The results showed that decreased knee joint forces and moments of both legs in ACLD patients were detected during the stance phase compared to the control group. Meanwhile, compared with ACLD knees, significantly fewer contact forces and flexion moments in ACLD combined with lateral and medial meniscal injury groups were found at the mid-stance, and ACLD with medial meniscal injury group showed a lower axial moment in the loading response ($p < 0.05$). In conclusion, ACLD knees exhibit reduced tibiofemoral joint forces and moments during jogging when compared with control knees. A combination of meniscus injuries in the ACLD-affected side exhibited abnormal kinetic alterations at the loading response and mid-stance phase.

Keywords: anterior cruciate ligament deficiency; meniscal injury; jogging; knee kinetics; inverse dynamics; musculoskeletal model

1. Introduction

Anterior cruciate ligament (ACL) and meniscus ruptures are common sports-related injuries and are often accompanied by a higher risk of osteoarthritis [1] even after ACL reconstruction [2,3]. The alterations in injured knee kinematics (such as joint angles and displacements), kinetics (such as joint moments, contact forces and ground reaction forces), and muscle activity are one of the reasons that lead to knee instability, thus causing secondary osteoarthritis. Numerous studies have investigated the characteristics of kinematics and kinetics in anterior cruciate ligament deficiency (ACLD) knees and found reductions in knee flexion moment and peak knee flexion angle, with patients adopting quadriceps avoidance [4] and stiffening strategy [5] gait patterns during level walking. For jogging, many studies focused on knee mechanics after ACL reconstruction. A study [6] reviewed the running biomechanics in patients with ACL reconstruction and pointed out that knee kinematics and kinetics were sensitive to sagittal plane motions. Asaeda et al. [7] investigated the relationship between knee muscle strength and knee biomechanics and found that both males and females were unable to restore their normal knee biomechanics from 3 months to 5 years. Moreover, some researchers [8–10] studied the knee mechanic alterations in ACLD patients during running and showed that kinematics changed both in the injured and uninjured knees. However, the effect of meniscus injuries on the ACLD knees during jogging is rarely reported.

On the other hand, accurate determination of kinetics involves difficulties and limitations in both in vitro cadavers and in vivo imaging studies. Musculoskeletal (MS) modeling could circumvent such shortcomings. Recently, many studies have performed multi-body dynamics modeling related to ACLD patients by using the musculoskeletal (MS) model and motion analysis system to estimate the joint and muscle forces. For instance, simulations were developed to predict ACL graft attachment locations during reconstruction [11], estimate knee contact mechanics and secondary kinematics with or without menisci [12], and evaluate the tibiofemoral joint forces of gait after ACL reconstruction [13]. Comparatively, there is little data on the evaluation of the kinetics of combined ACLD and meniscal injuries by using the MS model during jogging.

The aim of this study was to simulate the three-dimensional (3D) kinetics of ACLD knees with or without a medial or/and lateral meniscal injury during jogging by using the motion capture system and subject-specific multi-body dynamic model. The results could provide suggestions for clinical rehabilitation of muscle strength training and neuromuscular exercise. Our hypotheses were (1) the injured sides in ACLD knees would present significantly lower forces and moments during the stance phase, and the uninjured sides would be similar to healthy people; (2) ACLD combined with meniscal injury knees would show decreased moments during the stance phase compared with isolated ACLD knees.

2. Materials and Methods

2.1. Subjects

Twenty-nine young male patients with unilateral chronic ACLD knees (contralateral side intact) were recruited before undergoing ACL reconstruction. Ethical approval was obtained from the university's ethics committee and written informed consent was attained from all subjects. Twenty-three patients were injured through a non-contact accident, and 6 patients were through a contact accident. Most of the injury accidents occurred during basketball and football. Their activity level was evaluated by the Tegner score, which is a reliable and valid instrument to evaluate the activity level of patients with ACL injuries in the Chinese population [14]. The activity level of all the patients was normal before knee injuries (score range 3.0–6.0, Table 1). Among these patients, 12 patients had isolated unilateral ACL injuries (ACLD group), 5 had combined ACL and lateral meniscal injuries (ACLDL group), 5 had combined ACL and medial meniscal injuries (ACLDM group), and 7 had combined ACL and medial/lateral meniscal injuries (ACLDML group). The ACL injuries were documented by MRI and clinical examination and confirmed by arthroscopic findings, ranging from 6 months to 4 years prior to testing. Exclusion criteria were that patients had no previous ACL and concomitant meniscus and ligament ruptures, as well as no history of musculoskeletal pathologies of hip or ankle joints. Eleven lateral menisci were posterior longitudinal tears, and one was an anterior horn tear. All the medial meniscus were posterior complex horn tears. These patients had failed non-operative treatment of the injury and were therefore recognized as ACLD non-coppers. Fifteen healthy volunteered for this study (Control group). All the participants were male in this study because biomechanical characteristics were different between genders [9]. The morphological data are shown in Table 1. No significant differences were found between each of the two groups in terms of age, height, weight, body mass index (BMI) and pace. The time since injury for all the participants was over 6 months.

Table 1. Participant characteristics in each group *.

Parameters	Control	ACLD	ACLDL	ACLDM	ACLDML
Age (years)	29.33 ± 5.79	27.50 ± 1.98	26.60 ± 3.85	29.60 ± 5.41	27.71 ± 6.75
Height (cm)	171.83 ± 3.69	178.08 ± 8.28	179.80 ± 4.87	178.20 ± 7.09	180.29 ± 7.30
Weight (kg)	73.9 ± 6.75	82.58 ± 11.82	88.78 ± 23.46	85.70 ± 11.52	83.71 ± 14.37
BMI (kg/m^2)	25.09 ± 2.90	26.01 ± 2.90	27.26 ± 5.82	26.97 ± 3.01	25.64 ± 3.29
Pace(m/s)	2.33 ± 0.16	2.40 ± 0.22	2.47 ± 0.28	2.13 ± 0.28	2.42 ± 0.17
Tegner score	/	3.27 ± 0.75	4.60 ± 3.01	3.60 ± 1.02	5.33 ± 1.20
Time since injury (months)	/	10.50 ± 6.47	9.60 ± 3.29	15.40 ± 8.11	14.14 ± 5.37

* Data are presented as mean ± standard deviation. ACLD, anterior cruciate ligament deficiency; ACLDL, ACLD combined with lateral meniscal injury; ACLDM, ACLD combined with medial meniscal injury; ACLDML, ACLD combined with lateral and medial meniscal injury.

2.2. Data Collection and Modeling Analysis

Between January 2014 and December 2016, the experimental data was collected by an 8-camera motion capture system at a sample rate of 100 Hz (Vicon MX; Oxford Metrics, Yarnton, Oxfordshire, UK), where the marker trajectory data were filtered at 12 Hz. The ground reaction forces were collected by two embedded force plates at a sampling rate of 1000 Hz (AMTI, Advanced Mechanical Technology Inc., Watertown, MA, USA). A series of 14 mm markers were attached to the anatomical lower limb locations based on the plug-in-gait model to track the segmental motion during jogging (Figure 1). The participants were asked to run along a 10-m path at a self-selected speed where the kinematic data were recorded by eight cameras. No participants complained about pain during jogging. For each subject, five successful running trials were recorded, and these results were imported into the multi-body dynamics software AnyBody Modeling System (version 6.0.5, AnyBody™ Technology, Aalborg, Denmark) to estimate kinetics of the knee joint. The method of combining a motion capture system and musculoskeletal model in AnyBody Modeling System for gait analysis has been clinically validated in many studies, such as knee osteoarthritis [15,16], muscle activity [17,18] and kinematic characteristics of children's gait [19].

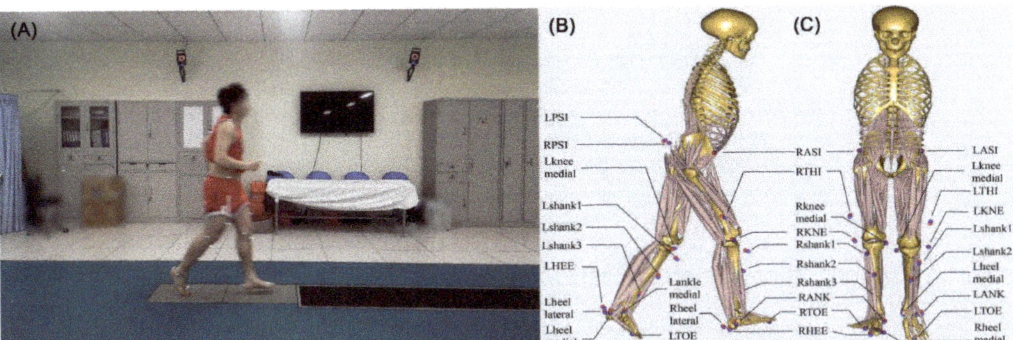

Figure 1. (**A**) Experimental data collection diagram; (**B**) The side view and (**C**) the front view of locations and names of the designated (red)/experimental (blue) markers on the musculoskeletal model when performing inverse dynamic analysis.

The Twente Lower Extremity Model [20] implemented in the AnyBody Modeling System was employed for the analysis. This model was comprised of 12 body segments, and 6 joint degrees of freedom were considered for each leg, with a spherical joint with 3 degrees of freedom for the hip joint and a universal joint with 2 degrees of freedom for the ankle joint. The knee joint was modeled as a hinge joint with 1 degree of freedom due to the soft tissue artifact error [21]. Based on the morphological parameters, each model was scaled with a mass-fat scaling algorithm to perform the subject-specific jogging simulation.

The min/max recruitment principle solver [22], which has better numerical convergence and physiological representation, was used to predict the muscle forces during the inverse dynamics analysis. The objective function minimizes the maximal muscle activity and can be formulated as follows:

$$\text{Minimize max}\left(\frac{f_i^{(M)}}{N_i}\right) \tag{1}$$

Subject to

$$\mathbf{Cf} = \mathbf{d}, \ 0 \leq f_i^{(M)} \leq N_i, \quad i \in \left\{1,\ldots,n^{(M)}\right\} \tag{2}$$

where $f_i^{(M)}$ and N_i refer to the muscle forces and the strength of the muscles, respectively. $\mathbf{f}$ contains muscle forces $\mathbf{f}^{(M)}$ and joint reactions $\mathbf{f}^{(R)}$. $\mathbf{C}$ is the coefficient-matrix and the right-hand side. $\mathbf{d}$ contains applied loads and inertia forces.

Five different running trials were simulated based on the experimental data, and the average values were used to perform kinetic analysis by using MATLAB (version: 2016b, MathWorks, Natick, MA, USA).

2.3. Statistics

The 3D tibiofemoral joint kinetics during jogging were normalized to 0–100% gait cycle at 1% intervals. Analysis of covariance with jogging speed as a covariate was performed among participant characteristics and kinetics points among the five groups. A priori sample calculation was conducted before statistical analyses by using the effect size Cohen's d. The forces and moments were analyzed by using a 1-way analysis of variance (ANOVA). A post hoc pairwise comparison using the Tukey multiple comparisons test was conducted when a significant difference was detected. For the interpatient kinetic differences of ACLD patients, two-way ANOVA was used to compare the tibiofemoral forces and moments among the ACLD-affected groups. The 2 within-patient factors were injury types (ACLD or ACLDL/ACLDM/ACLDML) and specific timings in the stance phase (10%, 20%, 30%, 40%, and 50% of gait cycle). The unpaired *t*-tests (independent sample) were performed between groups when the *F* ratio was at a significant difference level ($p < 0.05$).

3. Results

The jogging speed showed little effect on the kinetic parameters ($p > 0.05$). However, the injured and contralateral uninjured ACLDML knees showed large effect sizes in contact forces and moments compared to control knees (Table 2). Some gait parameters on the uninjured side showed large differences when there was a combined meniscal tear (ACLDL and ACLDM, d > 0.8 for ML, AP forces and axial moment). In particular, all the patients' knees exhibited large differences in the minimum flexion moment measure (Table 2). No large effect sizes except flexion moment were found between the ACLD and control groups.

Decreased knee joint forces and moments of both legs in ACLD patients were detected during the stance phase when compared to the control group, such as mediolateral force (Figure 2A,G), anteroposterior force (Figure 2C,I), axial moment (Figure 2D,J) and flexion moment (Figure 2F,L). In particular, the ACLD knees with lateral and medial meniscal deficiency, both for injured and uninjured knees, showed significant differences from the control knees at mid-stance as for the kinetic parameters measure (Table 3). All the ACLD patients, with or without meniscal injury, ran with lower extension moments than the control groups (Figure 2F,L). Meanwhile, significant differences in minimum flexion moment were observed both in injured knees (ACLD: -0.82 ± 0.46 Nm/(Bw $\times$ Bh); ACLDML: -0.68 ± 0.37 Nm/(Bw $\times$ Bh); $p < 0.05$) and uninjured knees (ACLDL: -0.85 ± 0.30 Nm/(Bw $\times$ Bh); ACLDM: -0.73 ± 0.21 Nm/(Bw $\times$ Bh); $p < 0.05$) during the mid-stance (control: -1.21 ± 0.21 Nm/(Bw $\times$ Bh)) (Table 3). No significant differences in parameter measure of joint reaction forces were found among control, ACLD, ACLDL, and ACLDM, except for the comparison of anteroposterior forces between control and uninjured knee in ACLDL patients (Table 3).

Table 2. The absolute value of Cohen's d effect size compared to the control group *.

	ACLD		ACLDL		ACLDM		ACLDML	
	Injured	Uninjured	Injured	Uninjured	Injured	Uninjured	Injured	Uninjured
Maximum ML force	0.18	0.49	0.70	**1.05**	0.50	**0.94**	**1.79**	**1.60**
Minimum PD force	0.42	0.34	0.39	0.61	0.31	0.67	**1.51**	**1.42**
Minimum AP force	0.44	0.59	**0.85**	**1.27**	0.59	**1.20**	**1.90**	**1.54**
Minimum axial moment	0.62	0.58	0.80	**1.05**	0.42	**1.17**	**1.45**	**1.00**
Maximum lateral moment	0.20	0.07	0.71	0.51	0.61	0.30	**1.52**	**1.64**
Minimum Flexion moment	**1.13**	**1.05**	**1.14**	**1.54**	**1.54**	**2.28**	**1.98**	**0.95**

* Small effect size: $0.2 \leq d < 0.5$; middle effect size: $0.5 \leq d < 0.8$; large effect size: $d > 0.8$ ACLD, anterior cruciate ligament deficiency; ACLDL, ACLD combined with lateral meniscal injury; ACLDM, ACLD combined with medial meniscal injury; ACLDML, ACLD combined with lateral and medial meniscal injury. ML, mediolateral; PD, proximodistal; AP, anteroposterior.

Table 3. Group Means (Standard Deviations) for kinetic parameters in the control limbs, injured and contralateral uninjured limbs during jogging.

	Control Group	Injured Groups							
		ACLD		ACLDL		ACLDM		ACLDML	
		Injured	Uninjured	Injured	Uninjured	Injured	Uninjured	Injured	Uninjured
Joint reaction forces/Bw									
Maximum ML force	2.40(0.43)	2.14(2.16)	2.17(0.52)	2.11(0.35)	1.97(0.33)	2.19(0.38)	1.97(0.54)	1.74(0.15) *	1.74(0.37) *
Minimum PD force	−8.60(1.30)	−7.94(1.87)	−8.10(1.65)	−8.10(1.16)	−7.86(0.88)	−8.21(1.04)	−7.64(1.83)	−6.90(0.54) *	−6.67(1.19) *
Minimum AP force	−7.97(2.36)	−6.48(4.34)	−6.53(2.46)	−6.04(1.90)	−5.09(1.92)*	−6.63(1.97)	−5.23(1.98)	−4.13(0.75) *	−4.65(1.53) *
Joint moment/(Bw × Bh) ($\times 10^{-1}$)									
Minimum axial moment	−0.68(0.26)	−0.49(0.36)	−0.53(0.25)	−0.49(0.13)	−0.41(0.25)	−0.58(0.14)	−0.40(0.14)	−0.32(0.22) *	−0.42(0.26)
Maximum lateral moment	0.60(0.14)	0.56(0.26)	0.59(0.13)	0.50(0.14)	0.53(0.13)	0.68(0.09)	0.56(0.11)	0.40(0.11) *	0.40(0.06) *
Minimum Flexion moment	−1.21(0.21)	−0.82(0.46)*	−0.98(0.23)	−0.95(0.28)	−0.85(0.30)*	−0.89(0.20)	−0.73(0.21) *	−0.68(0.37) *	−0.97(0.33)

* Statistically significant difference ($p < 0.05$) compared with controls from post-hoc pairwise comparison. ACLD, anterior cruciate ligament deficiency; ACLDL, ACLD combined with lateral meniscal injury; ACLDM, ACLD combined with medial meniscal injury; ACLDML, ACLD combined with lateral and medial meniscal injury. ML, mediolateral; PD, proximodistal; AP, anteroposterior; Bw, body weight; Bh, body height.

The result of interpatient differences with a significant F ratio (group difference: $p = 0.034$) showed that significant differences were observed at the mid-stance phase (20% of the gait cycle) between ACLD and ACLDML groups. The ACLDML knees presented significantly less contact force as well as less flexion moment compared to the ACLD groups. It was also noted that ACLDM knees showed lower axial moment compared with ACLD knees in the loading response (10% of the gait cycle) ($p < 0.05$). Kinetics in ACLDL knees did not show any significant difference compared to the other patient groups.

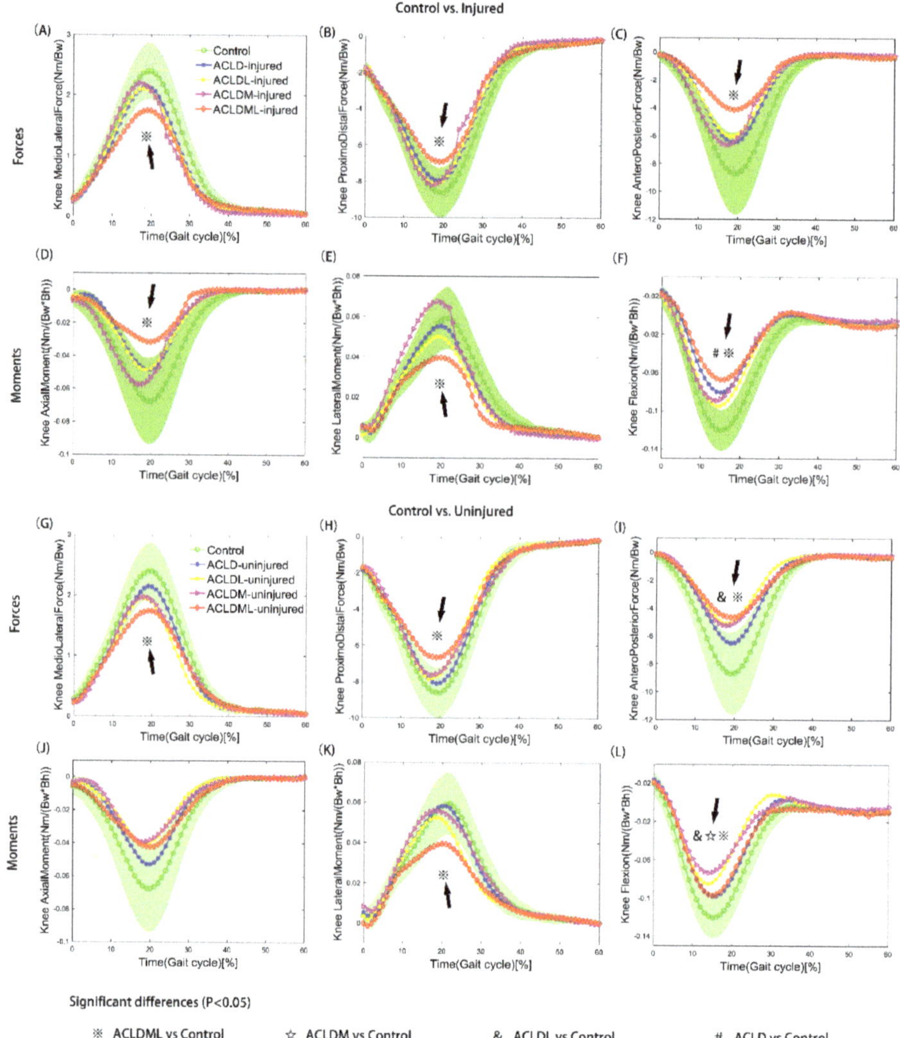

Figure 2. Tibiofemoral joint knee contact forces (**A**−**C**, **G**−**I**) and moments (**D**−**F**, **J**−**L**) of control and patient (ACLD, ACLDL, ACLDM and ACLDML) groups. Comparisons were conducted between control and injured limbs (**A**−**F**) as well as between control and uninjured limbs (**G**−**L**). Segments with significant statistical differences between every two groups were marked with symbols. The green shaded area represents the means ± standard deviation of the control group. ACLD, anterior cruciate ligament deficiency; ACLDL, ACLD combined with lateral meniscal injury; ACLDM, ACLD combined with medial meniscal injury; ACLDML, ACLD combined with lateral and medial meniscal injury. Bw: body weight; Bh: body height.

4. Discussion

This study investigated the kinetic abnormalities in male patients after ACLD combined with meniscal injuries during jogging. The results showed that forces and moments of uninjured side in ACLD combined with meniscus injury patients were lower compared to the control legs and were similar to those of the injured side during the stance of jogging trials. Therefore, this result partly supports our first hypothesis. Differences among these patient groups showed that the varying effects on the knee joint kinetics were dependent on the type of meniscal injuries. ACLD patients combined with higher severity degrees of meniscus tears (e.g., ACLDML vs. ACLDL) exhibited more abnormal knee joint contact forces and moments. These results supported our second hypothesis, which also confirmed previous findings that described the abnormal knee biomechanics after ACL and meniscal injuries during walking [16,23] and ascending stairs [24].

Some studies have found abnormal changes in ACLD [8] and reconstruction [3,6,7] knees during running. However, they did not report the alteration of contact forces, which is closely related to the development of osteoarthritis. A few works of literature investigated the knee contact mechanics involving the function of menisci [13] and ACL reconstruction [11,12] during walking by using MS modeling, but none have mentioned the biomechanical characteristics while jogging. As far as we know, this is the first study to estimate the kinetics of ACLD knees with concomitant medial or/and lateral meniscal tears during jogging by using the multi-body dynamics model. Based on our results, all the ACLD knees, either with or without concomitant meniscus injuries, present little discrepancy compared to healthy volunteers during the swing phase, which was in line with a systematic review [6]. Thus, this study focused on the different kinetic alterations during the stance phase of the gait cycle among control limbs and injured and uninjured limbs in ACLD combined with meniscal injuries.

Existing literature on the knee joint kinetic analysis during running mainly concentrated on isolated ACLD injuries, and the results were inconsistent due to different methodologies [8]. Although it was difficult to make a direct comparison with current studies, a similar trend towards decreased knee joint kinetics was detected both in contact forces [25] and moments [26] after ACL reconstruction. Meanwhile, the flexion moment measure was similar to the result described by Pratt et al. [26] when the moment peak was normalized to body weight between 10% and 50% of the stance phase (Pratt approximately 3.1 Nm/(Bw) vs. this work 2.1 Nm/(Bw)). The difference may result from the running speed, in which the velocity of the control group is lower in this study, 2.33 m/s, compared to the author's report of 4.24 m/s. The knee rotational moment was reported by some studies, which found a lower rotational moment in ACLD [8] and ACL reconstruction [27] knees compared to healthy limbs. The range of knee rotational moment for the control group compared favorably (0.07–0.13 Nm/(Bw × Bh) in [8] vs. 0.04–0.10 Nm/(Bw × Bh) in this study), and the peak value was similar to [27] when changing the unit Nm/(Bw·Bh) to Nm/(Bm·m) (body mass and leg length). Comparable kinetic results to ours have also been reported by Sasimontonkul et al. [28]. The joint reaction force on the tibia was calculated by the inverse dynamic method, and the average axial and shear force peaks were 8.0 Nm/(Bw) and 1.2 Nm/(Bw), respectively. In this study, the proximo-distal and mediolateral force peaks were 8.6 Nm/(Bw) and 2.4 Nm/(Bw), respectively. Different modeling methods may account for this discrepancy. In the literature [28], the model was composed of four segmented models and 21 simplified muscles, and the sum of the cubed muscle stresses was used to calculate muscle forces. In this study, the lower extremity model consists of 12 body segments and 55 muscle-tendon units, and the muscle recruitment principle solver minimizes the maximal muscle activity. All these comparisons confirm the correctness of this study, and our data provide additional findings in kinetic alterations of ACLD combined with different types of meniscal injury during jogging.

4.1. Kinetics of Injured and Uninjured Knees

The uninjured side of the ACLD patients, either with or without meniscal injury, presented similar knee contact forces and moments compared to the injured side. However, knee contact forces and moments kinetic measures of both injured and uninjured legs for ACLD patients were lower when compared to the control group. This result was consistent with some studies developed by ACLD and ACL reconstruction subjects. They reported that the subjects continued to present decreased knee moments after ACL injury and reconstruction [25,27] during early stance [29] and weight acceptance [30], even 6 and

12 months after ACL reconstruction [7]. However, there is limited evidence about the tibiofemoral joint contact forces in individuals with ACLD combined with meniscal injuries during jogging.

The kinetic alterations are often associated with the compensatory mechanisms in ACLD knees. Papadonikolakis et al. [31] summarized some in vivo biomechanical investigations and showed that ACLD patients adapted to ACL injury by different strategies over a long period of time. Despite disagreements over adaptation mechanisms by which type of ACLD patients adopt, in order to keep the knee joint stable during jogging, ACLD subjects tend to reduce the knee moments mainly through three ways: (1) increasing the hamstring activity, (2) decreasing gastrocnemius activity and (3) avoiding quadriceps contraction. Based on previous studies, it is likely that an adaptation strategy occurs due to a long time of repetitive activities after ACL injuries. The participants in this study all experienced a long time of adaptations since ACL and meniscal injuries (>6 months). Some kinetic changes in both limbs by the neural system have developed during the daily motions in order to restore functional gait. In our opinion, long-term loss of ACL and meniscus contributed to the decreased kinetics in the knee joint. The abnormal biomechanical [1,26] associated with muscle activity changes [7,27] in ACLD combined with meniscal injury patients increased the risk of osteoarthritis as well as the incidence of chondral and meniscus pathology. Previous studies have shown that joint unloading was associated with the cascade of early degenerative changes at the knee joint [1]. The lower compressive forces in ACLDML knees maybe result in a higher incidence of cartilage degeneration. The less knee flexion moment in ACLD combined with meniscal injury groups could be explained as a protective adaptation strategy to avoid excessive anterior tibial displacement. As the average maximum anterior displacement and external rotation in ACLD knees were significantly less than that of the contralateral healthy knee [32], lower flexion moment in ACLD combined with the meniscal injury group during the stance phase might also be an adaptation to the muscle firing patterns. It tends to reduce the flexion moment and increase the net quadriceps moment and the activity of the hamstrings to achieve a more normal tibiofemoral position during jogging. The lower flexion moment detected in ACLD, ACLDM, and ACLDML groups (Figure 2F, L) confirms this muscle co-contraction strategy. Subjects after ACL and meniscus injuries consistently stabilize their knee with a stiffening strategy [5,23] involving less knee flexion motion and higher muscle contraction.

From a clinical point of view, neuromuscular training for the ACLD, ACLDM, ACLDL, and ACLDML patients during jogging needs to be provided to restore normal knee biomechanics and thus reduce the risk for secondary cartilage injury and osteoarthritis. The ACLDML patients presented more alterations with the control group during jogging than the ACLDM and ACLDL patients. The ACLDML patients presented lower moments and contact forces in the sagittal, coronal and axial planes for both injured and uninjured legs. Therefore, ACLDML patients should enhance knee extension, abduction, and rotation muscle strength for both the injured and uninjured legs. The uninjured knees of the ACLDM and ACLDL patients showed lower rotation moments during jogging than the control group. The ACLDM and ACLDL patients should pay attention to the rotation muscle strength training of the uninjured legs. The injured knees of the ACLD patients presented lower rotation moments than control knees, which implies that the ACLD patients should enhance rotation muscle strength for the injured legs. Muscle strength and neuromuscular training should be conducted after ACL rupture, and normal joint kinetics should be one standard for returning to jogging.

4.2. Interpatient Differences of ACLD Knees with or without Meniscal Injury

Interpatient differences in ACLD knees, with or without meniscal injury, have been reported by many studies [16,23,24,33,34]. Similar to our group category, Zhang et al. performed two studies investigating kinematic characteristics during level walking [33] and ascending stairs [24]. Various effects of meniscus injuries on knee joint stability showed that a combined ACL and meniscal injuries could alter the biomechanical response in different ways, depending on the type of meniscal tears. Ren et al. [23] demonstrated a combination of "stiffening gait" and "pivot shift gait" patterns in ACLD knees with medial meniscus posterior horn tear, and this abnormal gait was confirmed by Liu et al. [16] in ACLD combined with medial and lateral meniscus injury patients. However, these studies mainly focused on the measurement of kinematics and kinetics during gait [16,23,33] and ascending stairs [24,34] rather than running.

Previous studies have shown that meniscal behavior showed characteristics of non-linearity, anisotropy, and non-homogeneity [35], resulting in complex kinetic alterations after meniscus injuries. In this study, we found that medial meniscus status was an important factor in the kinetic alterations of ACLD knees, as the ACLDML group presented significantly decreased kinetics at the mid-stance (20% of the gait cycle), and the ACLDM group showed fewer axial moments during the loading response (10% of the gait cycle) when comparing to ACLD group. Different functions of the meniscus may provide an explanation of the result. The medial meniscus was closely related to the sagittal tibial translation, while the lateral was more susceptible to transverse and frontal movements, playing an important role in postural stability [36]. Meanwhile, it was reported that most meniscal tears were peripheral posterior horn, and the peripheral meniscal tear in the medial meniscus contributed to the intense contraction of the hamstring muscles [37]. Medial meniscal injuries thus may result in more movement disorders when performing the simulation. However, further studies with more fundamental activities in ACLD combined with meniscal injury patients, such as pivoting [8] and cutting [10], are needed to confirm the effect of different medial and lateral meniscal injuries on the knee joint biomechanics.

Some limitations of this study should be noticed. First, the minimum number of subjects among these groups is small ($n = 5$), and it is not convincing to represent a whole population of individuals with different meniscus injury patterns during jogging. Second, individual movement alterations were evaluated only by kinetics, and the knee joint was modeled as a hinge joint due to soft tissue artifact error [21]. Muscle activity and small motions of kinematics, such as rotations in the transverse plane and translations in the coronal plane, were not considered in this study. Continued studies of a large number of subjects with a more comprehensive parameter analysis, for instance, muscle forces and ground reaction forces, will be necessary to confirm further the adaptation strategy in the patient who combined ACL and meniscus injuries before ACL reconstruction.

5. Conclusions

Using a motion capture system and subject-specific musculoskeletal models of lower limbs, this study first estimated tibiofemoral joint kinetic abnormalities in male patients of ACLD knees with or without a concomitant meniscal injury during jogging. The results demonstrated that a combined ACL and meniscal injury could alter kinetics of lower limbs during the stance phase depending on the presence and type of meniscal tears. Compared to the isolated ACL injuries, a combination of meniscus injuries in the ACLD-affected side would exhibit more abnormal alterations at the loading response phase (for the ACLDM group) and mid-stance phase (for the ACLDML group) during jogging. These abnormal alterations will result in the instability of the knee joint and increase the risk of degenerative changes and the occurrence of knee osteoarthritis. Our musculoskeletal model analysis of lower limbs also presented the necessity to implement musculoskeletal modeling of increasing complexity to reveal biomechanical characteristics and ultimately provide help for personalized rehabilitation schemes. Further studies should focus on the relationship between long-term biomechanical changes and the development of protective mechanisms after ACL and meniscal injuries, as well as subsequent specific treatment and rehabilitation.

Author Contributions: Conceptualization, Q.R. and Y.A.; methodology, S.R. and X.L. and H.L.; validation, Q.R., Z.M. and H.H.; formal analysis, Y.G. and Y.Z.; investigation, X.L., H.L. and Y.G.; resources, H.H. and S.R.; data curation, Z.L. and S.Z.; writing—original draft preparation, X.L. and S.R.; writing—review and editing, Q.R., X.L., S.R. and Z.M.; visualization, S.R.; supervision, Q.R., Y.A. and Z.M.; project administration, Q.R. and X.H.; funding acquisition, Y.A., Q.R., X.H. and Z.M. All authors have read and agreed to the published version of the manuscript.

Funding: This research was funded by the Natural Science Foundation of China Grant (Grant No: 11872074, 31900943, 12202412 and 12202413).

Institutional Review Board Statement: The study was conducted in accordance with the Declaration of Helsinki, and approved by the Ethics Committee of Peking University Third Hospital (IRB00006761-2012010) for studies involving humans.

Informed Consent Statement: Informed consent was obtained from all subjects involved in the study.

Data Availability Statement: Not applicable.

Conflicts of Interest: The authors declare no conflict of interest.

References

1. Wellsandt, E.; Gardinier, E.S.; Manal, K.; Axe, M.J.; Buchanan, T.; Snyder-Mackler, L. Decreased Knee Joint Loading Associated with Early Knee Osteoarthritis after Anterior Cruciate Ligament Injury. *Am. J. Sport. Med.* **2015**, *44*, 143–151. [CrossRef] [PubMed]
2. Barenius, B.; Ponzer, S.; Shalabi, A.; Bujak, R.; Norlén, L.; Eriksson, K. Increased Risk of Osteoarthritis after Anterior Cruciate Ligament Reconstruction: A 14-year follow-up study of a randomized controlled trial. *Am. J. Sport. Med.* **2014**, *42*, 1049–1057. [CrossRef]
3. Sigward, S.M.; Lin, P.; Pratt, K. Knee loading asymmetries during gait and running in early rehabilitation following anterior cruciate ligament reconstruction: A longitudinal study. *Clin. Biomech.* **2015**, *32*, 249–254. [CrossRef]
4. Berchuck, M.; Andriacchi, T.P.; Bach, B.R.; Reider, B. Gait adaptations by patients who have a deficient anterior cruciate ligament. *J. Bone Jt. Surg.* **1990**, *72*, 871–877. [CrossRef]
5. Hurd, W.J.; Snyder-Mackler, L. Knee instability after acute ACL rupture affects movement patterns during the mid-stance phase of gait. *J. Orthop. Res.* **2007**, *25*, 1369–1377. [CrossRef] [PubMed]
6. Pairot-De-Fontenay, B.; Willy, R.W.; Elias, A.R.C.; Mizner, R.L.; Dubé, M.-O.; Roy, J.-S. Running Biomechanics in Individuals with Anterior Cruciate Ligament Reconstruction: A Systematic Review. *Sport. Med.* **2019**, *49*, 1411–1424. [CrossRef]
7. Asaeda, M.; Deie, M.; Kono, Y.; Mikami, Y.; Kimura, H.; Adachi, N. The relationship between knee muscle strength and knee biomechanics during running at 6 and 12 months after anterior cruciate ligament reconstruction. *Asia-Pac. J. Sport. Med. Arthrosc. Rehabil. Technol.* **2019**, *16*, 14–18. [CrossRef]
8. Bohn, M.B.; Petersen, A.K.; Nielsen, D.B.; Sørensen, H.; Lind, M. Three-dimensional kinematic and kinetic analysis of knee rotational stability in ACL-deficient patients during walking, running and pivoting. *J. Exp. Orthop.* **2016**, *3*, 27. [CrossRef] [PubMed]
9. Théoret, D.; Lamontagne, M. Study on three-dimensional kinematics and electromyography of ACL deficient knee participants wearing a functional knee brace during running. *Knee Surg. Sport. Traumatol. Arthrosc.* **2006**, *14*, 555–563. [CrossRef]
10. Waite, J.C.; Beard, D.J.; Dodd, C.A.F.; Murray, D.W.; Gill, H.S. In vivo kinematics of the ACL-deficient limb during running and cutting. *Knee Surg. Sport. Traumatol. Arthrosc.* **2005**, *13*, 377–384. [CrossRef]
11. Koo, Y.-J.; Jung, Y.; Seon, J.K.; Koo, S. Anatomical ACL Reconstruction can Restore the Natural Knee Kinematics than Isometric ACL Reconstruction during the Stance Phase of Walking. *Int. J. Precis. Eng. Manuf.* **2020**, *21*, 1127–1134. [CrossRef]
12. Sanford, B.A.; Williams, J.L.; Zucker-Levin, A.R.; Mihalko, W.M. Tibiofemoral Joint Forces during the Stance Phase of Gait after ACL Reconstruction. *Open J. Biophys.* **2013**, *03*, 277–284. [CrossRef]
13. Hu, J.; Xin, H.; Chen, Z.; Zhang, Q.; Peng, Y.; Jin, Z. The role of menisci in knee contact mechanics and secondary kinematics during human walking. *Clin. Biomech.* **2019**, *61*, 58–63. [CrossRef]
14. Huang, H.; Zhang, D.; Jiang, Y.; Yang, J.; Feng, T.; Gong, X.; Wang, J.; Ao, Y. Translation, Validation and Cross-Cultural Adaptation of a Simplified-Chinese Version of the Tegner Activity Score in Chinese Patients with Anterior Cruciate Ligament Injury. *PLoS ONE* **2016**, *11*, e0155463. [CrossRef] [PubMed]
15. Wong, C.; Bencke, J.; Rasmussen, J. Triceps surae strength balancing as a management option for early-stage knee osteoarthritis: A patient case. *Clin. Biomech.* **2022**, *95*, 105651. [CrossRef]
16. Liu, X.; Huang, H.; Yin, W.; Ren, S.; Rong, Q.; Ao, Y. Anterior cruciate ligament deficiency combined with lateral and/or medial meniscal injury results in abnormal kinematics and kinetics during level walking. *Proc. Inst. Mech. Eng. Part H J. Eng. Med.* **2019**, *234*, 91–99. [CrossRef]
17. Huang, H.; Yin, W.; Ren, S.; Yu, Y.; Zhang, S.; Rong, Q.; Ao, Y. Muscular Force Patterns during Level Walking in ACL-Deficient Patients with a Concomitant Medial Meniscus Tear. *Appl. Bionics Biomech.* **2019**, *2019*, 7921785. [CrossRef]

18. Dupré, T.; Dietzsch, M.; Komnik, I.; Potthast, W.; David, S. Agreement of measured and calculated muscle activity during highly dynamic movements modelled with a spherical knee joint. *J. Biomech.* **2019**, *84*, 73–80. [CrossRef] [PubMed]
19. Ziziene, J.; Daunoraviciene, K.; Juskeniene, G.; Raistenskis, J. Comparison of kinematic parameters of children gait obtained by inverse and direct models. *PLoS ONE* **2022**, *17*, e0270423. [CrossRef] [PubMed]
20. Horsman, M.K.; Koopman, H.; van der Helm, F.; Prosé, L.P.; Veeger, H. Morphological muscle and joint parameters for musculoskeletal modelling of the lower extremity. *Clin. Biomech.* **2007**, *22*, 239–247. [CrossRef]
21. Andersen, M.S.; Benoit, D.L.; Damsgaard, M.; Ramsey, D.K.; Rasmussen, J. Do kinematic models reduce the effects of soft tissue artefacts in skin marker-based motion analysis? An in vivo study of knee kinematics. *J. Biomech.* **2010**, *43*, 268–273. [CrossRef]
22. Marra, M.A.; Vanheule, V.; Fluit, R.; Koopman, B.H.F.J.M.; Rasmussen, J.; Verdonschot, N.; Andersen, M.S. A Subject-Specific Musculoskeletal Modeling Framework to Predict In Vivo Mechanics of Total Knee Arthroplasty. *J. Biomech. Eng.* **2015**, *137*, 020904. [CrossRef]
23. Ren, S.; Yu, Y.; Shi, H.; Miao, X.; Jiang, Y.; Liang, Z.; Hu, X.; Huang, H.; Ao, Y. Three dimensional knee kinematics and kinetics in ACL-deficient patients with and without medial meniscus posterior horn tear during level walking. *Gait Posture* **2018**, *66*, 26–31. [CrossRef] [PubMed]
24. Zhang, Y.; Huang, W.-H.; Ma, L.; Lin, Z.; Huang, H.; Xia, H. Kinematic characteristics of anterior cruciate ligament deficient knees with concomitant meniscus deficiency during ascending stairs. *J. Sport. Sci.* **2016**, *35*, 402–409. [CrossRef] [PubMed]
25. Bowersock, C.D.; Willy, R.W.; DeVita, P.; Willson, J.D. Reduced step length reduces knee joint contact forces during running following anterior cruciate ligament reconstruction but does not alter inter-limb asymmetry. *Clin. Biomech.* **2017**, *43*, 79–85. [CrossRef] [PubMed]
26. Pratt, K.A.; Sigward, S.M. Knee Loading Deficits during Dynamic Tasks in Individuals Following Anterior Cruciate Ligament Reconstruction. *J. Orthop. Sport. Phys. Ther.* **2017**, *47*, 411–419. [CrossRef]
27. Lewek, M.; Rudolph, K.; Axe, M.; Snyder-Mackler, L. The effect of insufficient quadriceps strength on gait after anterior cruciate ligament reconstruction. *Clin. Biomech.* **2002**, *17*, 56–63. [CrossRef]
28. Sasimontonkul, S.; Bay, B.K.; Pavol, M.J. Bone contact forces on the distal tibia during the stance phase of running. *J. Biomech.* **2007**, *40*, 3503–3509. [CrossRef]
29. DeVita, P.; Hortobagyi, T.; Barrier, J.; Torry, M.; Glover, K.L.; Speroni, D.L.; Money, J.; Mahar, M.T. Gait adaptations before and after anterior cruciate ligament reconstruction surgery. *Med. Sci. Sport. Exerc.* **1997**, *29*, 853–859. [CrossRef]
30. Timoney, J.M.; Inman, W.S.; Quesada, P.M.; Sharkey, P.F.; Barrack, R.L.; Skinner, H.B.; Alexander, A.H. Return of normal gait patterns after anterior cruciate ligament reconstruction. *Am. J. Sport. Med.* **1993**, *21*, 887–889. [CrossRef]
31. Papadonikolakis, A.; Cooper, L.; Stergiou, N.; Georgoulis, A.D.; Soucacos, P.N. Compensatory mechanisms in anterior cruciate ligament deficiency. *Knee Surg. Sport. Traumatol. Arthrosc.* **2003**, *11*, 235–243. [CrossRef] [PubMed]
32. Andriacchi, T.P.; Dyrby, C.O. Interactions between kinematics and loading during walking for the normal and ACL deficient knee. *J. Biomech.* **2005**, *38*, 293–298. [CrossRef] [PubMed]
33. Zhang, Y.; Huang, W.-H.; Yao, Z.; Ma, L.; Lin, Z.; Wang, S.; Huang, H. Anterior Cruciate Ligament Injuries Alter the Kinematics of Knees with or without Meniscal Deficiency. *Am. J. Sport. Med.* **2016**, *44*, 3132–3139. [CrossRef] [PubMed]
34. Hosseini, A.; Li, J.-S.; Gill, I.T.J.; Li, G. Meniscus Injuries Alter the Kinematics of Knees with Anterior Cruciate Ligament Deficiency. *Orthop. J. Sport. Med.* **2014**, *2*. [CrossRef]
35. Ferroni, M.; Belgio, B.; Peretti, G.; Di Giancamillo, A.; Boschetti, F. Evolution of Meniscal Biomechanical Properties with Growth: An Experimental and Numerical Study. *Bioengineering* **2021**, *8*, 70. [CrossRef] [PubMed]
36. Lee, J.-H.; Heo, J.-W.; Lee, D.-H. Comparative postural stability in patients with lateral meniscus versus medial meniscus tears. *Knee* **2018**, *25*, 256–261. [CrossRef] [PubMed]
37. Smith, J.P., III; Barrett, G.R. Medial and Lateral Meniscal Tear Patterns in Anterior Cruciate Ligament-Deficient Knees: A prospective analysis of 575 tears. *Am. J. Sport. Med.* **2001**, *29*, 415–419. [CrossRef]

bioengineering

Article

Soccer Scoring Techniques—A Biomechanical Re-Conception of Time and Space for Innovations in Soccer Research and Coaching

Gongbing Shan [1,2,*] and Xiang Zhang [2]

[1] Biomechanics Lab, Faculty of Arts & Science, University of Lethbridge, Lethbridge, AB T1K 3M4, Canada
[2] Department of Physical Education, Xinzhou Teachers' University, Xinzhou 034000, China; zhangxiang@xztu.edu.cn
* Correspondence: g.shan@uleth.ca; Tel.: +1-403-329-2683

Abstract: Background: Scientifically, both temporal and spatial variables must be examined when developing programs for training various soccer scoring techniques (SSTs). Unfortunately, previous studies on soccer goals have overwhelmingly focused on the development of goal-scoring opportunities or game analysis in elite soccer, leaving the consideration of player-centered temporal-spatial aspects of SSTs mostly neglected. Consequently, there is a scientific gap in the current scoring-opportunity identification and a dearth of scientific concepts for developing SST training in elite soccer. Objectives: This study aims to bridge the gap by introducing effective/proprioceptive shooting volume and a temporal aspect linked to this volume. Method: the SSTs found in FIFA Puskás Award (132 nominated goals between 2009 and 2021) were quantified by using biomechanical modeling and anthropometry. Results: This study found that players' effective/proprioceptive shooting volume could be sevenfold that of normal practice in current coaching. Conclusion: The overlooked SSTs in research and training practice are commonly airborne and/or acrobatic, which are perceived as high-risk and low-reward. Relying on athletes' talent to improvise on these complex skills can hardly be considered a viable learning/training strategy. Future research should focus on developing player-centered temporal-spatial SST training to help demystify the effectiveness of proprioceptive shooting volume and increase scoring opportunities in soccer.

Keywords: proprioceptive shooting volume; zero possession shot; scoring opportunity identification; airborne; biomechanical modeling; anthropometry

Citation: Shan, G.; Zhang, X. Soccer Scoring Techniques—A Biomechanical Re-Conception of Time and Space for Innovations in Soccer Research and Coaching. *Bioengineering* **2022**, *9*, 333. https://doi.org/10.3390/bioengineering9080333

Academic Editors: Rui Zhang and Wei-Hsun Tai

Received: 2 July 2022
Accepted: 22 July 2022
Published: 23 July 2022

Publisher's Note: MDPI stays neutral with regard to jurisdictional claims in published maps and institutional affiliations.

1. Introduction

Soccer is the most popular sport in the World. Based on the information from Fédération Internationale de Football Association (FIFA), the game is played and watched on five continents with 265 million players and 4 billion fans, i.e., over 50% of the world population (7.7 billion) are linked to the game [1–3]. Yet, contrary to the popularity of the game, the number of scientific inquiries on key motor control skills, i.e., soccer scoring techniques (SSTs), appears disproportionately low when compared to the participation-to-scientific study ratios of other sports skills, such as complex gymnastics skills [4]. As a result, the scientific understanding of SSTs lags far behind its practice, with most participants acquiring various SSTs through individual experience rather than science-based instruction [5,6]. To make the matter worse, there is a dearth of scientific investigation on the many complex SSTs, such as the jumping turning kick, e.g., a nominated goal for the FIFA Puskás Award 2019, performed by Ibrahimović [7] and the diving scorpion kick, which won the FIFA Puskás Award 2017 [8]. These SSTs seem virtuosic in scope and are commonly believed to be the talent ability solely of soccer stars [9]. Obviously, relying on the aptness of the athletes to improve these virtuosic SSTs can hardly be considered a viable learning or coaching strategy. Worst of all, the terms surrounding SSTs used in practice and research have been confusing [10]; researchers and practitioners are not sure how many SSTs are available for the game. This scenario hinders not only the scientific studies on many exceptional SSTs but also the development of novel coaching methods for learning these SSTs.

It is well known that the greatest attraction of soccer is goal scoring. Compared to many other sports, goals are relatively rare in soccer–on average, less than three goals per game in FIFA world cups since the 1960s [11]. Because of their rarity, soccer goals are extremely exciting for millions of fans. The various means by which gameplay moves toward a goal can be thought of as an improvised drama, where emotional tension is built over long periods only to be fully released when the goal is achieved. This characteristic contributes to making soccer the most popular spectator sport in the world. Therefore, an essential core of soccer research and coaching is to help athletes master aesthetically eye-catching SSTs for increasing both scoring possibilities and the excitement of the game.

1.1. Novel Scientific Studies Required for the World's Most-Popular Sport

Through systematic identification, a recent study [10] has revealed that there are 43 SSTs that exist in current soccer games. Surprisingly, there are only a handful of SSTs documented in the existing training/coaching literature and, consequently, the comprehensive list of SST training is presently limited to maximal instep kick (including the curled kick), jumping headers, and volley/side volleys [12–14]. Since the existing scientific understanding could not supply enough help to practitioners in developing training methods for learning more SSTs, practical *modi operandi* have been the main driving force for keeping SST exuberant.

At the practical level, celebrations of the goal of the month, of the season, of the year, and of the decade, are widely adopted in professional soccer leagues for recognizing players who are deemed to have scored the "most beautiful goal" [10]. Those glorious goals are normally chosen by a combination of panel experts and a public vote. This practice has been successfully used to encourage and promote aesthetically significant goals from players, as it sparks the creativity of the athletes to develop novel/unique SSTs through "self-learning". Even infrequently, under this practical *modi operandi*, exceptional SSTs have been developed by a few talented athletes from time to time. These virtuosic skills have become models for more athletes and coaches to impersonate and duplicate blindly, i.e., learning without insight knowledge obtained from research. Especially the brilliant goals scored in soccer's flagship tournaments, such as those held by FIFA, UEFA (Union of European Football Associations) and other professional national soccer leagues with an international reputation (e.g., La Liga/Spain, Serie A/Italy, Bundesliga/Germany, Premier League/England, and Ligue 1/French), are mimicked by worldwide soccer players. Biomechanically, there are two issues related to such blind self-learning: learning efficiency and injury risk.

Soccer shots, like many complicated human movements, are trained motor skills. Studies have demonstrated that systematic and scientific inquiry into the biomechanics of human motor skills has great potential to demystify complicated human motor skills for efficient learning [15–18]. Hence, knowledge obtained from biomechanics studies can help optimize performance outcomes while simultaneously reducing the risk of training-related injury. There are two useful biomechanical analyses: kinematic and kinetic quantification of motor skills. In terms of motor learning, kinematics has significant utility in terms of improving teaching and learning, while an understanding of kinetics is essential for reducing the risk of learning and playing-related injury. Both kinematics and kinetics have significant utility in raising practitioners' awareness of biomechanical "cause and effect", thus giving them additional tools and knowledge to optimize their practice [6,17,19–21].

From a motor learning point of view, one common criterion of the honored goals in soccer's flagship tournaments is their repeatability, e.g., the FIFA Puskás Award states that "the goal should not be the result of luck or mistakes by the other team" [22], suggesting that the nominated goals and the related SSTs are theoretically repeatable, i.e., entrainable. What is needed for establishing a scientific training system is the knowledge of the biomechanical parameters that influence the quality of the soccer shots. Unfortunately, most of the 43 SSTs [10] identified so far are "off the radar" to researchers. The SSTs overlooked in research are normally airborne and/or acrobatic, perceived as high-risk and low-reward. Counting on athletes' talents to improvise on those SSTs is neither scientific nor realistic. Hence, novel biomechanical studies are needed for the scientific discovery of these airborne and/or acrobatic SSTs. As a result, the skills that are virtuosic in appearance may be eminently trainable and less a product of the improvisatory abilities of individual players.

1.2. Literature Review

Web of Science is a reputable resource due to its guaranteed proofed scientific content [23]. Among all types of papers, review articles draw upon published articles to provide a great overview of the existing literature on a topic. As such, to reveal the current state of scientific studies on soccer scoring with rigorous and quality information, the Web of Science database was searched for systematic review articles in December 2021. The search was performed by using the keywords "football" and "soccer", with each associated with the terms "goal analysis" and "review". The initial search identified 31 papers in the database. The titles and abstracts of 31 articles were then screened according to their relevance to SSTs in elite soccer, resulting in 27 studies being eliminated. At the end of the screening procedure, four systematic review articles [24–27] between February 2018 and January 2020 received further in-depth reading to identify the current research state in this area. The articles show that soccer scoring quantification started as early as 1968, investigating the statistical relationship between the number of passes and goals scored by analyzing 3213 professional matches from 1953 to 1968 [28]. Over the past half-century, more and more notational measurements were developed and applied to finding the factors influencing soccer goal scoring. These measurements include total passing frequencies [29], ball possession time [30,31], shots at goal [32,33], the position of an attempt on goal (e.g., three longitudinal areas: right, center, and left, and zone: ultra-defensive, defensive, central, offensive, and ultra-offensive) [34,35], the influence of set plays or dead ball routines [36,37] etc. In short, the previous studies on soccer goals have overwhelmingly focused on the development of goal-scoring opportunities or game performance analysis, i.e., tactical strategies, and overlooked the role of various SSTs on soccer goals.

As matter of a fact, very few studies have analyzed the special kicking skills that elite goal scorers possess [9,27]. As revealed by the recent systematic review article, the existing research studies have not fully explored the final actions of the players in goal situations [27]. There is no doubt that the current studies can provide important references for coaches in designing training programs for developing the tactical strategies of a team. However, they could not improve the SST training. It is well known that all tactical strategies learned and trained aim at achieving the ultimate shot opportunity, and the key to the success or failure of the shot depends largely on the SST selected by the player.

A study on defense in all 306 German Bundesliga games from the 2010/2011 season [38] showed that the top teams have a faster defensive reaction time compared to the remaining teams. The result suggests that the time taken to perform a shot is becoming shorter and shorter as the competition level increases. FIFA has vividly described this development trend as "every nanosecond is special" [39]. As a consequence, more and more airborne shots are seen in current elite games, and the height of an airborne kick is increasing higher and higher in the top elite-level games. Unfortunately, both current research and existing training systems can no longer keep up with actual real-game development.

1.3. The Current State of SST Knowledge

In this vein, few existing studies have found that shots with the feet seem to achieve between 70–80% of the goals, whereas the rest of the goals are achieved from headers [40,41]. More details have been added by a recent study [10], finding that there are 43 SSTs identified in elite soccer games, and over 60% of them are airborne SSTs. The airborne shots contribute to over 50% of goals, with shots attempted by foot, head, or chest. These findings indicate that scoring opportunity identification has to consider factors linked to airborne shots. Sadly, the most up-to-date systematic review article on the biomechanics of kicking in soccer is a dozen-years old, which focuses only on the instep kick [42], with a dearth of scientific studies on most of the airborne SSTs [4,43].

There are also limited studies on the anatomic parts of the foot used to shoot. One study has found that the instep seems to be the most used anatomic part, followed by the inside part of the foot [44]. The recent study conducted by Zhang and Shan [10] added more details to this; depending on the scorer's body posture when shooting (e.g., facing, side-facing, or back-facing) and ball position (e.g., ground or airborne), the instep, dorsi-side, inside, outside, toe, heel, or plantar-side are selected by scorers for shots.

Due to the limited knowledge related to SSTs, there is a scientific gap in current scoring-opportunity identification. From a scientific standpoint, both temporal and spatial variables must be examined when evaluating scoring opportunities. In essence, scoring chances are an issue of maximizing the probabilities of scoring, which cannot be determined without basic theories and knowledge related to the temporal-spatial identification and quantification of a player's possibilities to shoot.

From the temporal perspective, it is well known that even if in possession of a free ball, a player will likely not be free for long; defenders will attempt to thwart the shot. A representative study [40], which analyzed all goals in the English Premier League during the 2008/2009 season, found that scoring with zero possession (i.e., a "one-touch shot" where a player shoots as a ball is passing by) accounts for 69.3% of goals. Setting up a shot with one or more contacts of the ball results in only 17.9% of goals. These results indicate that the longer a player possesses the ball, the lower the scoring chance. The least favorable situation for scoring is when players score after individual dribbling (12.8%).

Regarding spatial variables, existing studies have shown that 65–90% of goals are scored when an attacking player possesses the ball in or near the opponents' penalty area, and he/she is not hindered by defenders, i.e., a "free ball" [44–47]. The studies have actually confirmed the practitioners' empirical evidence. This explains why most of the current goal-scoring training overwhelmingly emphasizes the geographic location of the ball on the playing field.

In brief, the present research scheme fails to consider the player-centered temporal-spatial aspects of SSTs and underemphasizes air-attack (3D consideration) related to shooting techniques.

1.4. Research Aims and Research Questions

The current scenario would suggest that ground-breaking research is needed in order to develop science-based SST training regimes for improving scoring possibility. The current paper aims to lay a foundation for launching a ground-breaking study via the re-conception of temporal and spatial factors related to soccer shooting, identifying elements that could be applied in the entrainment of complex SSTs via biomechanical quantification. The study has three specific research questions to be answered (i.e., the main goals of the current study):

- Scientifically, what is the content of temporal–spatial opportunity identification related to SST?
- What is a quantitative and reliable method to evaluate the improvement of scoring possibility through science-based training?
- What is the theoretical implication of these findings related to the innovative development of SST training?

2. Materials and Methods

For answering these research questions, the current study initiates a novel theoretical framework, which has its origin in elite soccer. Three steps are involved in the establishment of the original framework: (1) selection of temporal criteria for efficiency recognition, (2) identification of new/potential spatial variables via the video-based analysis of all 132 nominated goals of the FIFA Puskás Award between 2009 and 2021 [22], and (3) quantification of scoring possibility via biomechanical modeling.

2.1. A New Theoretical Framework for Developing SST Training—Focusing on Time in Space

Scoring opportunity can be defined by the feasibility of shooting successfully. Practically, it is well known that when an attacker gets a scoring opportunity, his/her chance to shoot the ball typically does not last very long. In these brief moments, a player needs to shoot the ball quickly (the temporal aspect of goal scoring) and accurately (the spatial feature of goal scoring). The reality in elite games is actually more complicated than this simplified thinking. Regardless of the geographic opportunities one could gain, only considering the dynamic relationship between an attacker, the goal, and the 3D position of the ball related to the scoring chance identification, is not enough to reach a decision without knowing the biomechanical characteristics of various SSTs. Therefore, a new framework should link the temporal and spatial details to the SSTs for ameliorating the scoring possibility.

Since there is no existing framework for considering the temporal and spatial aspects of SSTs simultaneously, our study will bridge the gap by creating a new theoretical framework. The original framework covers factors related to the temporal efficiency and spatial effectiveness of SSTs. The basic ideas of the framework are: (1) to mathematically evaluate the effective shooting volume for scoring chance quantification, i.e., a quantitative determination of spatial feasibility for shooting, and (2) for any dynamic ball (chance) covered by the effective shooting volume, to choose a proper SST for shooting without delay, i.e., temporal efficiency of shooting. The novel aspect of the original framework is to introduce 3D considerations (of player-centered temporal-spatial consideration) for the training of SSTs in order to improve scoring possibility.

The key element of the framework is the effective shooting volume. Its quantification will inevitably require the study of the overlooked airborne and/or acrobatic SSTs. These SSTs are intricate, belonging to gymnastic-like motor skills. Previous studies have revealed that the learning quality of gymnastic-like motor skills depends on one's proprioceptive ability and can be progressively developed through structured repetitive training [48–51]. Hence the effective shooting volume can also be named as the player's proprioceptive shooting volume. This volume determination would fundamentally inform if a dynamic ball should be counted as a goal chance (i.e., the ball falls within or passes through the volume) or not.

In summary, the core of the novel framework is to create a new quantitative way for (1) enlarging the effective shooting volume to cover more goal chances and (2), at the same time, enhancing the shooting ability to ensure one-touch shots within this volume. Based on these core elements, this original framework can be denominated as "Focusing On Time In Space". It aims to nexus the temporal efficiency and spatial effectiveness of maximizing soccer scoring possibilities. Whence, it would build a scientific foundation for innovations in future SST research and training.

2.2. Selection of Temporal Criteria

Our selection of temporal criteria is based on a study related to the temporal efficiency of SSTs carried out by Durlik and Bieniek (2014) [40]. Since players scoring after individual dribbling involves other motor skills rather than SST skills only, a modification has been made, i.e., the scoring after individual dribbling was excluded. Additionally, any further actions before shooting have been proven to decrease scoring possibility [40]; therefore, it is logical to select the amount of possession before shooting so as to recognize overall temporal efficiency. Hence, the following categories were selected for the quantification in this study:

- one-touch-shot (zero-possession), where a player shoots as a ball is passing by;
- One possession, i.e., setting the ball and then shooting;

- the other ball control strategies combined, i.e., 2+ possession maneuvers.

The temporal criteria were evaluated via the statistical results of the 132 Puskás nominated goals between 2009 and 2021 [22].

2.3. Identification of New Spatial Variables

Regarding spatial analysis, previous studies have focused only on field geography [44–47] and have neglected those factors related to SSTs, e.g., the athlete's body orientation facing, side-facing, or back-facing the goal, and the spatial position of the ball at the instance of a shot [10]. To give more detail to the latter aspect, the spatial position of the ball has horizontal and vertical components. Horizontally, using the goal and the player as positional references, the ball can be between them, beyond them, or to the side of them. Vertically, the ball can be airborne or on the ground (Figure 1). Therefore, in the current study, these new spatial parameters related to SSTs at the instance of a shot were induced and designated for quantitative analysis.

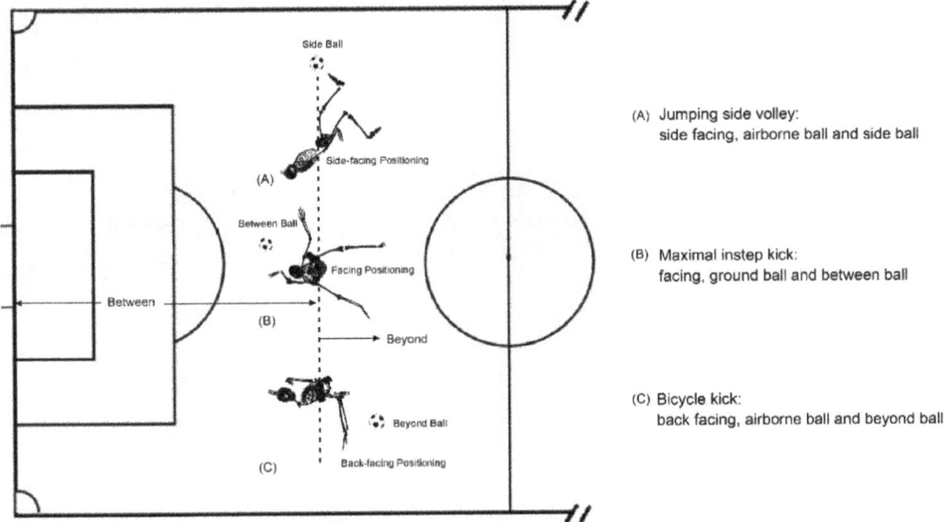

Figure 1. The identification and clarification of spatial factors. (**A**) The 3D quantification of a jumping side volley–top view [4,43]. (**B**) The 3D quantification of the maximal instep kick–top view [52–54]. (**C**) The 3D quantification of the bicycle kick–top view [9,43].

Combinatorically, the above selected temporal and spatial parameters were applied to the quantitative examination of all 132 Puskás shooting videos. Descriptive statistics (i.e., pie chart) were applied to summarize these categorical data (i.e., percentage distribution) in order to reveal the characteristics of the selected parameters and their contributions to scoring goals at the top elite level.

2.4. Quantification of the Proprioceptive Shooting Volume via Biomechanical Modeling

Dimensional data obtained from 3D motion capture and/or full-body biomechanical modeling were used to estimate the proprioceptive/effective shooting volume. The quantification works as follows:

- Apply 3D biomechanical modeling to quantify various kicking techniques [4,9,53,55];
- Target and select three SSTs to determine the anterior–posterior, medial–lateral, and vertical dimensions when the kicks are being performed. The selection of SSTs represents the shooting ability that an athlete can obtain after the current training methods or that talented elite players have already performed in elite soccer;

- Use the three dimensions of the selected SSTs to estimate the proprioceptive shooting volumes of the two different performance levels and to determine the difference between them.

The 3D dimensional data used in this study are either from the authors' previous 3D quantification studies or gained through both anthropometrical study [56] and biomechanical model estimation/simulation. Clearly, different selections of SSTs will result in distinct sizes of the volume, i.e., various SSTs will lead to a differentiation in scoring possibility. Logically, the larger one's effective shooting volume, the more goal chances one will possess. Through the novel framework, the current article has, for the first time, visualized the concept of how to establish a link between the temporal-spatial aspects of SSTs and goal chance quantification. This conceptual exploration is currently missing in soccer research and practice.

3. Results

The analysis results of temporal and spatial factors related to goal-chance identification are shown in Figure 2. Compared to the outcomes of temporal efficiency from the previous study [40], i.e., 69.3% of goals scored for zero-possession shots, 17.9% of goals scored after two or more possessions, and 12.8% of goals scored for shots after dribbling, the temporal branch of Figure 1 based on the 132 FIFA Puskás Award nominated goals shows comparable results: 56.8% (0 possession), 70.5% (0 & 1 possession), and 29.5% (2 + possessions), respectively.

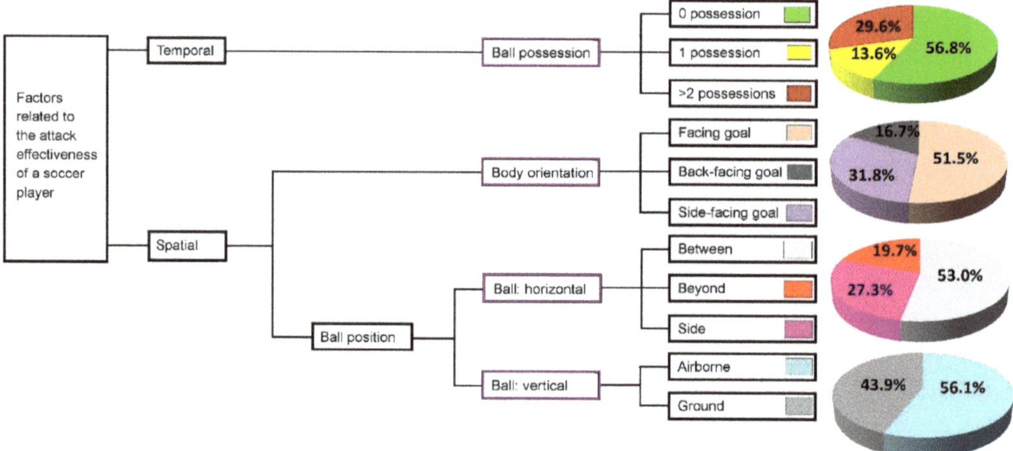

Figure 2. The temporal and spatial factors related to goal chance identification and their contributions to goals: evidence drawn from all 132 FIFA Puskás Award nominees' goals, 2009–2021.

The results of the spatial analysis of the 132 Puskás goals denoted that, as a ball comes into an athlete's proprioceptive shooting volume, a player's body orientation can be facing, side-facing, or back-facing (Figures 1 and 2) relative to the position of the goal. In terms of body orientation, at the moment of shooting among the 132 Puskás goals, about half (51.5%) were achieved facing the goal, while 31.8% were "side-facing", and 16.7% were "back-facing" (Figure 2). The statistics for horizontal ball position are comparable, with 53.0% of goals having occurred when the ball was between the player and goal, while side and beyond balls accounted for the remaining 47.0%. Regarding vertical ball positioning, 56.1% of the goals were airborne, and only 43.9% were ground balls.

The three-dimension comparisons between the SSTs trained by the current system and the selected SSTs performed by elite players are shown in Figure 3. As elaborated in the method section, these three-dimensional values were applied to quantitatively estimate the shooting volume formed by skills in terms of a player's body height (BH). It should be noted that (1) the estimation represents the maximum volume one could reach, and (2) the dimensions normalized by BH increased the generalization of the method [55]. In current soccer coaching practice, the systematically trained SSTs are the maximal instep kick (including the curled kick), jumping headers, and volleys and side volleys [12–14]. Therefore, the proprioceptive shooting volume of the currently-trained SSTs could be calculated by the following dimensions:

- Anterior–posterior dimension is 1.3 BH, i.e., the last-stride length of the maximal instep kick [52,54,55] (Figure 3A);
- Medial–lateral dimension is 0.8 BH, i.e., the lateral reach of the side volley of 0.4 BH by each leg (Figure 3B);
- Vertical dimension of 1.4 BH, i.e., the jumping height for a header (Figure 3C).

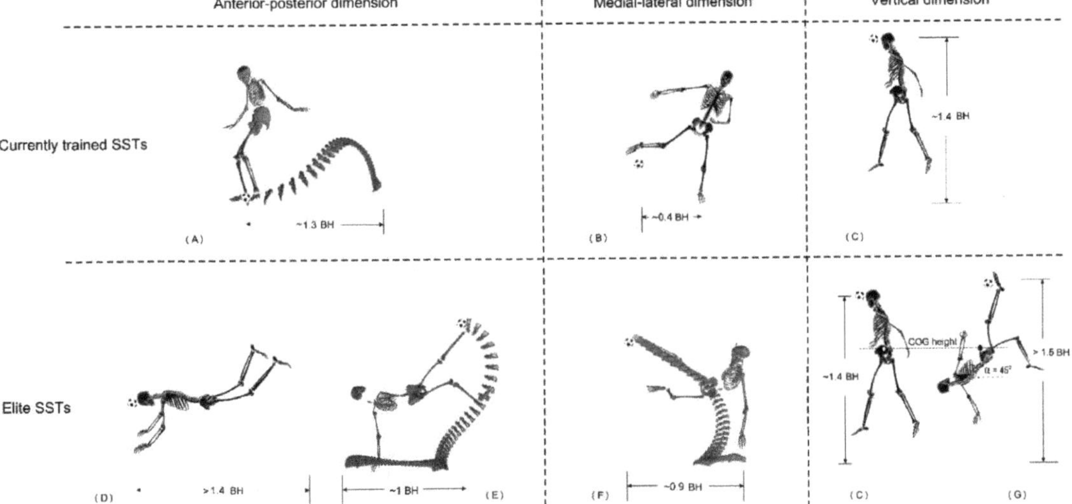

Figure 3. The dimensions of selected SSTs and their influences on the quantification of the attack space (scoring chance): (**A**) The 3D quantification of the maximal instep kick–side view with timely trace of the kicking foot, obtained via 3D-motion analysis and biomechanical modeling [52–54]. (**B**) Lateral dimension of volley kick–distance estimated using biomechanical modeling and anthropometrical study [5,56]. (**C**) Vertical dimension of a jumping header–distance estimated using jump biomechanics [57,58] and anthropometrical modeling [56]. (**D**) Frontal dimension of a diving header–distance estimated using diving biomechanics [59] and anthropometrical modeling [56]. (**E**) The 3D quantification of the bicycle kick–side view with timely trace of the kicking foot, obtained via 3D-motion analysis and biomechanical modeling [9,43,56]. (**F**) The 3D quantification of jumping side volley–frontal view with timely trace of the kicking foot, obtained via 3D-motion analysis and biomechanical modeling [4,43]. (**G**) Potential attack height reached by using a bicycle kick at different trunk-angle orientations–biomechanical model estimation [9,56].

In the case of a 1.8 m tall athlete, when the above dimensions are considered, the calculation of his/her maximum effective shooting volume would be 2.34 m × 2.52 m × 1.44 m, or very roughly 8.5 m³.

A similar quantitative estimation was also applied to the selected SSTs performed by elite players. The practical potential of the quantification of the proprioceptive/effective shooting volume is shown in Table 1. For the same case (1.8 m tall player), if the volumes of two acrobatic SSTs (i.e., bicycle kick and jumping side volley, Figure 3E,F, performed by talented elite soccer players) are included in the calculation, the same player's effective shooting volume would be 4.14 m × 2.52 m × 3.24 m, roughly 35.4 m^3, or approximately four times the shooting volume of the normally practiced techniques in current training practice. If additional SSTs, such as the diving header (Figure 3D) and an improved bicycle kick (trunk angled 45°, in addition to the current 0° bicycle kick, Figure 3G) are included, the effective shooting volume increases to 4.32 m × 2.70 m × 5.04 m, roughly 58.8 m^3, or approximately seven times the normally practiced shooting volume. Ad hoc, more quantitative estimations could be performed, e.g., the long-jump header [10] performed by Cristiano Ronaldo in the 2008/2009 UCL season (Figure 4). If this SST could be quantified, the effective shooting volume would definitely increase further.

Table 1. The comparison of the proprioceptive/effective shooting volume from the current practice (100%) to the player-centered temporal-spatial training of airborne/acrobatic SSTs.

Trained through Currently Practice	Achieved by Talented Elite Athletes	Theoretical Potential
100%	~400%	~700%

Figure 4. The long-jump header (the player is identified in red circle on the two left frames). Currently, there is no study available for revealing the 3D dimension of this SST (the figure was generated from the video of UEFA TV video, published in 2021 [60]).

4. Discussion

Due to the rarity of goal chances in soccer, it is highly relevant to establish a theoretical system for innovating research and training practice around goal scoring. Obviously, the re-conception of scoring chance quantification and its relationship to the temporal and spatial aspects of SSTs would have great potential to form a foundation on which novel developments of various SSTs could be built. Unfortunately, scientific studies on the influence of temporal-spatial factors on shooting effectively (i.e., quantify the scoring chance and turning that scoring chance into a goal) have not yet been conducted. This is why our study proposed a novel framework for this area. Based on the results, this section makes an effort to answer research questions 1–3 to elaborate the conceptual metaphors of the novel framework.

4.1. Temporal-Spatial Opportunity Identification

The evidence gained from the 132 FIFA Puskás goals has clearly indicated that a new theoretical framework is needed for re-examining and/or allocating the known and unknown factors related to existing SST, and breakthrough studies are desirable to quantify the influences of those factors on temporal-spatial opportunity identification.

4.1.1. Temporal Efficiency

The first challenge would be a temporal enumeration, i.e., the number of possessions of the ball during preparation for shooting. As portrayed before, even if in possession of a "free ball", a player will likely not be free for long; defenders will attempt to thwart the shot. Therefore, any delay in preparing a shot will decrease temporal efficiency. The results summarized in this study clearly indicate that a sudden attack would create the highest temporal efficiency for converting a scoring chance into a goal. This result is comparable to the representative study [40] on the analysis of all goals from the English Premier League during the 2008/2009 season. Yet, the temporal aspect does not stand alone, and obviously, it interacts with spatial factors.

4.1.2. Spatial Effectiveness

The special aspect presents a more challenging concept, but one thing is certain: it must consist of more than mere field geography. A player's proprioceptive abilities influence goal-opportunity identification, i.e., as the soccer ball comes into a player's proprioceptive shooting volume, his/her ability to perform a shot under various body orientations with the 3D spatial positions of the ball must fall within these abilities in order for him/her to attempt to score. Current knowledge shows that proprioceptive abilities can be developed through repetitive training through structured and targeted training [48–51].

Yet, spatial factors may be over-simplified in the present training practice. There seems to be an existing research and coaching emphasis on shots taken (1) facing the goal and (2) with the ball between the player and the goal [12–14]. It is not surprising that most training and practice regimes concentrate on the variables that account for the higher percentages of goals found in the pie charts of Figure 2. However, this neglects a significant percentage of the scoring opportunities that may exist in smaller "slices of the pie". Some of these might be improved through scientifically structured training. For example, Swedish player Ibrahimovic's bicycle kick against England in 2013 (i.e., the Puskás award 2013 [61]) was a goal achieved as he ran away from, and not toward, the goal. This shot can be characterized as follows: 0 possession, back facing the goal, a beyond ball, and an air attack. Currently, there is no science-based training program available to practitioners for learning this skill [9,43]. Therefore, the skill is generally considered a product of the improvisatory and virtuosic abilities of an individual player, not a trainable one. The question arises whether or not the training regimes that develop the SST shooting components required for such "virtuosity" might improve scoring percentage by increasing a player's proprioceptive/effective shooting volume. The 132 Puskás goals have shown that 16.7% of the goals were scored by using back-facing goal techniques, and 19.7% of the goals were scored with the ball located beyond (not between) the goal and the players (Figure 2). These "unusual" goals would signify that what is required is a new theoretical framework to logically link the temporal and spatial factors for a systematical exploration of these acrobatic skills for developing science-based training methods.

4.1.3. Lost in Time and Space in the Current Scoring Research

The current study reveals that joint consideration of the temporal and spatial aspects of scoring has been mostly neglected in the current SST research and coaching literature, i.e., scientific studies on SST could be considered lost in time and space. The results of Puskás goals imply that a few talented elite athletes, via years of practice, have magically linked temporal efficiency and spatial effectiveness in developing improvisatory abilities. Such abilities can turn "impossible" goals (commonly identified as non-chance goals) into goals. The scientific re-conception considered in this study would suggest that the biomechanical quantification of the temporal–spatial factors of these extraordinary SSTs could help us demystify these "magic kicks", and evidently identify a scoring chance as "yes" or "no" instead of "impossible". Without the scientific quantification of these temporal–spatial challenges, the identification of a scoring chances is vague, subjective, and unclear. Unfortunately, this is the situation for the current studies on soccer scoring, especially for the spatial challenges related to SSTs.

In summary, the current study is the one that has attempted to originate a novel framework for scientifically pinpointing the content of temporal–spatial opportunity identification. As such, future innovations in SST research and training system development could be launched by applying the proposed framework.

4.2. Quantification of Athletes' Proprioceptive/Effective Shooting Volume—A Key for Scoring-Opportunity Identification

Scientifically, the quantification of a player's attack volume would impartially show whether or not a ball should be counted as a chance of a goal. At the present time, there are few, if any, studies on the training manipulation and expansion of a player's effective shooting volume. Within the current knowledge, the training effect on the volume depends on the learner's spatial motor control ability, i.e., the proprioceptive competence, which is highly entrainable [48–51]. Up to now, this volume is still limited by the following SSTs in coaching practice [12–14]:

- Maximal instep kick (including curled kicks), characterized as kicks facing the goal, between balls, and ground balls;
- Headers, characterized as shots facing/side-facing the goal, between/side balls, and an air attack;
- Volleys, characterized as kicks facing/side-facing the goal, between/side balls, and an air attack below the hip.

Therefore, the current training system would under develop a player's proprioceptive/effective shooting volume, reducing their possibilities to shoot and decreasing their goal-scoring chances.

The results of this study clearly demonstrated that the limited proprioceptive shooting volume is a result of the insufficient SSTs with which a player has been entrained. The results of Table 1 would suggest that talented athletes, e.g., Cristiano Ronaldo and Zlatan Ibrahimović (both winners of the Puskás Award and are able to perform a bicycle kick and a jumping side volley), have found ways to master unique acrobatic SSTs for increasing their proprioceptive/effective shooting volume by four times that of normally trained players. Evidently, these unusual SSTs contribute to enlarging their proprioceptive shooting volume. Furthermore, this study has revealed that, theoretically, the increased volume could potentially still be enlarged to seven or more times that of the "normally" practiced shooting volume (Table 1). The ramifications of our results are far-reaching. Even if less than half of these theoretical gains could be realized through the training of the underutilized "slices of the pie" from Figure 1, a player's effective proprioceptive shooting volume could be doubled or tripled.

In short, to maximize scoring probabilities, it is essential to intensely expand the dimensions of the proprioceptive shooting volume, which is heavily influenced by the SSTs available to an athlete. The goals of Puskás nominees unveil that there are more than a dozen SSTs that should be considered as potential skills to maximize a player's proprioceptive shooting volume. Based on FIFA criteria [22], these SSTs are undeniably repeatable; as such, they should be entrainable. What we need are scientific studies that demystify these SSTs and establish new and effective training regimes for developing these extraordinary SSTs among young and future players.

4.3. Focusing on Time in Space—The Nexus for Uniting Time Efficiency and Spatial Effectiveness

The nexus for uniting the time efficiency and spatial effectiveness of players should be rooted in the development strategy of future SST training. The logical sequence would be:

- Firstly, master as many SSTs as possible through science-based SST training. This step would aim at enlarging the effective 3D dimensions of the proprioceptive shooting volume, i.e., increasing spatial effectiveness. This is the foundation for maximizing scoring possibilities;
- Secondly, entraining for decision ability in properly selecting an SST for an accurate attack by means of shooting various dynamic balls. This step would focus on temporal efficiency. Mastering more SSTs would allow a player to choose the proper SST for ensuring a zero-possession shot, regardless of the ball's horizontal and vertical position as well as an attacker's dynamic posture (Figures 1–3);
- Lastly, further increase the decision ability with pre-judgements via practical training (e.g., game-like situations). This step would concentrate on making an achievable

decision based on a reasonable pre-judgement during an in-game situation. After mastering various SSTs, a player would have more options to choose from when preparing an attack. How to plan an attack to preclude defensive players (for a clear shot) would be vital for shooting success. An excellent example is the scorpion kick (one of 2014 Puskás nominated goals [62]) performed by Ibrahimovic. During the game between Paris Saint-Germain and SC Bastia (Ligue 1/French), Ibrahimovic was running across the opponent's penalty area (side-facing the goal) and received a chance on goal via his teammate's airborne pass, which was falling into his proprioceptive shooting volume. Two defenders were applying a one-on-one defense on each of his sides (Figure 5 left) and attempting to defend the chance. Using his perfect pre-judgement ability, Ibrahimovic purposely moved further to let the ball fell behind him. This planned action precluded the defenders and created a free ball for a successful scorpion kick (Figure 5 right). Such a decision clearly needs the player to be able to perform a kick with a "beyond" and "airborne" ball.

Figure 5. The scorpion kick, a nominated goal of FIFA Puskás Award 2014 performed by Ibrahimović (the figure was generated from the video of FIFA Puskás Award 2014).

Regrettably, scientific studies, as well as coaching practice, fall far behind in initiating the above training regime. Most of the identified SSTs (currently 43 in total [10]) cannot be systematically trained due to the lack of scientific investigation and understanding. As elaborated before, the SSTs overlooked by researchers and the current training regime are the airborne and/or acrobatic attack techniques, such as the scorpion kick, diving scorpion kick, jumping side volley, jumping turning kick, long-jump turning header, diving header, sliding kick, rabona kick, and more [10]. The fatal effect of the overlook is the decrease in time efficiency and spatial effectiveness.

One common characteristic of the SSTs overlooked in research and coaching practice is exceedingly complex, labeled as high-risk and low-reward [9]. These skills, for most athletes, cannot be learned without insightful guidance. It is time for researchers to conduct quantitative studies for (1) demystifying the complexity of the skills, (2) identifying the skills required for various ball positions and dynamic body postures, and (3) develop training programs for injury-free exercise, since the accurate performance of the airborne and/or acrobatic SSTs requires repetitive training.

This is the first study on the time and space aspects of SSTs. It is understandable that there are limitations associated with this study. There are two obvious ones. First, due to the unavailability of the 3D data of most SSTs, the current quantification on the spatial effectiveness can only be used as a reference for practitioners. More future studies are inevitably needed for reaching an accurate result. Second, there might be gender-based control-pattern variations. A quantitation of female athletes, as well as comparisons between the males and the females, must be conducted in the future studies.

Summarized above, the practical information for coaches and players is as follows: (1) the more SSTs an athlete can perform, the more goal chances he will have; and (2) the more airborne SSTs an athlete can master, the more efficient and effective his shooting attacks will be. Yet, merely relying on the aptness of an athlete to master these extraordinary SSTs would be hit or miss. Such an approach cannot be considered as a viable coaching strategy. A more science-structured learning, based on the re-conceptual organization of temporal-spatial aspects of SSTs should be developed. The development ought to firstly focus on improving players' spatial awareness in terms of body orientation and ball spatial position and, subsequently, increase the temporal efficiency through practices of a one-touch-shot within this improved volume. In short, focusing on Time-in-Space would be the nexus for uniting time efficiency and spatial effectiveness in future SSTs learning and training.

5. Conclusions

Retrospectively, from a scientific standpoint, both temporal and spatial variables must be examined when evaluating soccer scoring opportunities. Unfortunately, field geography (i.e., the development of goal scoring opportunities) or game analysis in elite soccer tends to dominate the attention of researchers and practitioners and consideration of the player-centered temporal-spatial aspects of SSTs is mostly neglected.

Goal scoring research fails to address the time and space related to SSTs. Space certainly consists of more than mere field geography. A player's trained SSTs influence both scoring opportunity identification and the dimensions of his/her attack space. The development of novel training programs should, first, focus on the increase of the proprioceptive shooting volume through mastering as many SSTs as possible; then, it should concentrate on the training of selecting a proper SST to reach one-touch-shots within the enlarged attack volume, i.e., focusing on time in space.

The current study reveals that a player has to learn airborne and acrobatic SSTs in order to increase his/her spatial effectiveness, as well as the temporal efficiency of shooting. Therefore, scientific studies are indisputably needed to demystify the complexity of these skills for developing their learning and training.

The great attraction of soccer for millions of fans is the goal. Various techniques for scoring goals are sources of excitement. More frequent use of airborne and acrobatic SSTs for goals can only enhance the excitement of the game. Therefore, this new theoretical framework would bring more excitement by promoting novel studies and developing innovative training programs for learning and practicing various SSTs.

Author Contributions: Both G.S. and X.Z. are actively involved in the conceptualization, methodology, validation, formal analysis, investigation, resources, writing—original draft preparation, review and editing, as well as visualization of the paper. G.S. is for the funding acquisition. All authors have read and agreed to the published version of the manuscript.

Funding: This research was funded by Discovery Development Grant of National Sciences and Engineering Research Council of Canada (NSERC), Grant #: DDG-2021-00021.

Institutional Review Board Statement: Not applicable.

Informed Consent Statement: Not applicable.

Data Availability Statement: Not applicable.

Conflicts of Interest: The authors declare no conflict of interest.

References

1. FIFA. 265 Million Playing Football. 2007. Available online: https://www.fifa.com/mm/document/fifafacts/bcoffsurv/emaga_9384_10704.pdf (accessed on 6 February 2018).
2. Sawe, B.E. The Most Popular Sports in the World. 2018. Available online: https://www.worldatlas.com/articles/what-are-the-most-popular-sports-in-the-world.html (accessed on 16 June 2018).
3. Worldometers. Current World Population. 2018. Available online: http://www.worldometers.info/world-population/ (accessed on 11 July 2018).
4. Zhang, X.; Shan, G.; Liu, F.; Yu, Y. Jumping Side Volley in Soccer—A Biomechanical Preliminary Study on the Flying Kick and Its Coaching Know-How for Practitioners. *Appl. Sci.* **2020**, *10*, 4785. [CrossRef]
5. Shan, G.; Zhang, X. From 2D leg kinematics to 3D full-body biomechanics-the past, present and future of scientific analysis of maximal instep kick in soccer. *Sports Med. Arthrosc. Rehabil. Ther. Technol.* **2011**, *3*, 23. [CrossRef] [PubMed]
6. Shan, G.; Zhang, X.; Wan, B.; Yu, D.; Wilde, B.; Visentin, P. Biomechanics of coaching maximal instep soccer kick for practitioners. *Interdiscip. Sci. Rev.* **2019**, *44*, 12–20. [CrossRef]
7. FIFATV. FIFA PUSKAS Award 2019 Nominee: Zlatan Ibrahimovic. 2019. Available online: https://www.youtube.com/watch?v=oV6L2cNr0MU (accessed on 11 June 2019).
8. FIFATV. Olivier GIROUD—FIFA PUSKAS Award 2017 Final. 2017. Available online: https://www.youtube.com/watch?v=wGOhxt5Ksc0 (accessed on 11 June 2020).
9. Shan, G.; Visentin, P.; Zhang, X.; Hao, W.; Yu, D. Bicycle kick in soccer: Is the virtuosity systematically entrainable? *Sci. Bull.* **2015**, *60*, 819–821. [CrossRef]
10. Zhang, X.; Shan, G.; Liu, F.; Yang, S.; Meng, M. Diversity of Scoring, Ingenuity of Striking, Art of Flying—Conceptual and Systematical Identification of Soccer Scoring Techniques. *Phys. Act. Rev.* **2021**, *9*, 86–99. [CrossRef]
11. Statista. Average Number of Goals Scored per Game at the FIFA World Cup from 1930 to 2018. 2020. Available online: https://www.statista.com/statistics/269031/goals-scored-per-game-at-the-fifa-world-cup-since-1930/ (accessed on 26 January 2020).
12. Bénézet, J.-M.; Hasler, H. *Youth Football*; Galledia AG: Berneck, Switzerland, 2017.
13. FIFA. FIFA Coaching Manual. 2017. Available online: https://premiersoccerinstitute.com/wp-content/uploads/2017/04/Fifa-Coaching-manual.pdf (accessed on 17 July 2018).
14. Thömmes, F. *Fußballtraining für jeden Tag: Die 365 besten Übungen; [+ 16 Seiten moderne Entwicklungstendenzen]*; Stiebner Verlag GmbH: Grünwald, Germany, 2012.
15. Ballreich, R.; Baumann, W. *Grundlagen der Biomechanik des Sports (The Basics of Biomechanics in Sports)*; Enke Verlag: Stuttgart, Germany, 1996.
16. Liu, Y.; Kong, J.; Wang, X.; Shan, G. Biomechanical analysis of Yang's spear turning-stab technique in Chinese martial arts. *Phys. Act. Rev.* **2020**, *8*, 16–22. [CrossRef]
17. Visentin, P.; Shan, G.; Wasiak, E.B. Informing music teaching and learning using movement analysis technology. *Int. J. Music Educ.* **2008**, *26*, 73–87. [CrossRef]
18. Yu, D.; Yu, Y.; Wilde, B.; Shan, G. Biomechanical characteristics of the axe kick in Tae Kwon-Do. *Arch. Budo* **2012**, *8*, 213–218. [CrossRef]
19. Turner, C.; Visentin, P.; Oye, D.; Rathwell, S.; Shan, G. Pursuing Artful Movement Science in Music Performance: Single Subject Motor Analysis With Two Elite Pianists. *Percept. Motor Skills* **2021**, *128*, 1252–1274. [CrossRef]
20. Wąsik, J.; Shan, G. Factors influencing the effectiveness of axe kick in taekwon-do. *Arch. Budo* **2014**, *10*, 29–34.
21. Zhang, Z.; Li, S.; Wan, B.; Visentin, P.; Jiang, Q.; Dyck, M.; Li, H.; Shan, G. The influence of X-factor (trunk rotation) and experience on the quality of the badminton forehand smash. *J. Hum. Kinet.* **2016**, *53*, 9–22. [CrossRef] [PubMed]
22. FIFA. FIFA Introduces New FIFA Puskás award to Honour the "Goal of the Year". 2009. Available online: https://www.fifa.com/mensyoutholympic/news/y=2009/m=10/news=fifa-introduces-new-fifa-puskas-award-honour-the-goal-the-year-1120531.html (accessed on 16 January 2016).
23. Gusenbauer, M.; Haddaway, N.R. Which academic search systems are suitable for systematic reviews or meta-analyses? Evaluating retrieval qualities of Google Scholar, PubMed, and 26 other resources. *Res. Synth. Methods* **2020**, *11*, 181–217. [CrossRef] [PubMed]
24. Memmert, D.; Rein, R. Match analysis, big data and tactics: Current trends in elite soccer. *German J. Sports Med./Dtsch. Z. Sportmed.* **2018**, *69*, 65–72. [CrossRef]
25. Pratas, J.M.; Volossovitch, A.; Carita, A.I. Goal scoring in elite male football: A systematic review. *J. Hum. Sport Exerc.* **2018**, *13*, 218–230. [CrossRef]
26. Sarmento, H.; Clemente, F.M.; Araújo, D.; Davids, K.; McRobert, A.; Figueiredo, A. What performance analysts need to know about research trends in association football (2012–2016): A systematic review. *Sports Med.* **2018**, *48*, 799–836. [CrossRef]
27. Rodenas, J.G.; Malaves, R.A.; Desantes, A.T.; Ramirez, E.S.; Hervas, J.C.; Malaves, R.A. Past, present and future of goal scoring analysis in professional soccer. *Retos: Nuevas Tend. Educ. Fís. Deporte Recreación* **2020**, *37*, 774–785.
28. Reep, C.; Benjamin, B. Skill and chance in association football. *J. R. Stat. Soc.* **1968**, *131*, 581–585. [CrossRef]
29. Hughes, M.; Franks, I. Analysis of passing sequences, shots and goals in soccer. *J. Sports Sci.* **2005**, *23*, 509–514. [CrossRef]
30. Jones, P.; James, N.; Mellalieu, S.D. Possession as a performance indicator in soccer. *Int. J. Perform. Anal. Sport* **2004**, *4*, 98–102. [CrossRef]

31. Lago-Peñas, C.; Dellal, A. Ball possession strategies in elite soccer according to the evolution of the match-score: The influence of situational variables. *J. Hum. Kinet.* **2010**, *25*, 93–100. [CrossRef]

32. Moura, F.A.; Van Emmerik, R.E.A.; Santana, J.E.; Martins, L.E.B.; de Barros, R.M.L.; Cunha, S.A. Coordination analysis of players' distribution in football using cross-correlation and vector coding techniques. *J. Sports Sci.* **2016**, *34*, 2224–2232. [CrossRef] [PubMed]

33. Pratas, J.M.; Volossovitch, A.; Carita, A.I. The effect of performance indicators on the time the first goal is scored in football matches. *Int. J. Perform. Anal. Sport* **2016**, *16*, 347–354. [CrossRef]

34. Camerino, O.F.; Chaverri, J.; Anguera, M.T.; Jonsson, G.K. Dynamics of the game in soccer: Detection of T-patterns. *Eur. J. Sport Sci.* **2012**, *12*, 216–224. [CrossRef]

35. Zurloni, V.; Cavalera, C.; Diana, B.; Elia, M.; Jonsson, G. Detecting regularities in soccer dynamics: A T-pattern approach. *Rev. Psicol. Deporte* **2014**, *23*, 157–164.

36. Bangsbo, J.; Peitersen, B. *Soccer Systems and Strategies*; Human Kinetics: Winsor, Canada, 2000.

37. Yiannakos, A.; Armatas, V. Evaluation of the goal scoring patterns in European Championship in Portugal 2004. *Int. J. Perform. Anal. Sport* **2006**, *6*, 178–188. [CrossRef]

38. Vogelbein, M.; Nopp, S.; Hökelmann, A. Defensive transition in soccer–are prompt possession regains a measure of success? A quantitative analysis of German Fußball-Bundesliga 2010/2011. *J. Sports Sci.* **2014**, *32*, 1076–1083. [CrossRef]

39. FIFATV. Every Nanosecond Is Special. 2021. Available online: https://www.youtube.com/shorts/yGg3ff2iyd4 (accessed on 16 January 2022).

40. Durlik, K.; Bieniek, P. Analysis of goals and assists diversity in English Premier League. *J. Health Sci.* **2014**, *4*, 47–56.

41. Armatas, V.; Mitrotasios, M. Analysis of goal scoring patterns in the 2012 European Football Championship. *Sport J.* **2014**, *1*, 1–11.

42. Lees, A.; Asai, T.; Andersen, T.B.; Nunome, H.; Sterzing, T. The biomechanics of kicking in soccer: A review. *J. Sports Sci.* **2010**, *28*, 805–817. [CrossRef]

43. Shan, G. Biomechanical Know-how of Fascinating Soccer-kicking Skills—3D, Full-body Demystification of Maximal Instep Kick, Bicycle kick & Side Volley. In Proceedings of the 8th International Scientific Conference on Kinesiology, Opatija, Croatia, 10–14 May 2017; University of Zagreb: Zagreb, Croatia, 2017.

44. Michailidis, Y.; Michailidis, C.; Primpa, E. Analysis of goals scored in European Championship 2012. *J. Hum. Sport Exerc.* **2013**, *8*, 367–375. [CrossRef]

45. Seabra, F.; Dantas, L.E. Space definition for match analysis in soccer. *Int. J. Perform. Anal. Sport* **2006**, *6*, 97–113. [CrossRef]

46. Wright, C.; Atkins, S.; Polman, R.; Jones, B. Factors associated with goals and goal scoring opportunities in professional soccer. *Int. J. Perform. Anal. Sport* **2011**, *11*, 438–449. [CrossRef]

47. Armatas, V.; Yiannakos, A. Analysis and evaluation of goals scored in 2006 World Cup. *J. Sport Health Res.* **2010**, *2*, 119–128.

48. Alsmith, A.J.; Longo, M.R. Where exactly am I? Self-location judgements distribute between head and torso. *Conscious. Cognit.* **2014**, *24*, 70–74. [CrossRef] [PubMed]

49. Aman, J.E.; Elangovan, N.; Yeh, I.-L.; Ekonczak, J. The effectiveness of proprioceptive training for improving motor function: A systematic review. *Front. Hum. Neurosci.* **2015**, *8*, 1075. [CrossRef]

50. Sherrington, C.S. On the proprioceptive system, especially in its reflex aspect. *Brain* **1907**, *29*, 467–482. [CrossRef]

51. Wong, J.D.; Kistemaker, D.A.; Chin, A.; Gribble, P.L. Can proprioceptive training improve motor learning? *J. Neurophys.* **2012**, *108*, 3313–3321. [CrossRef]

52. Shan, G. Influence of gender and experience on the maximal instep soccer kick. *Eur. J. Sport Sci.* **2009**, *9*, 107–114. [CrossRef]

53. Shan, G.; Daniels, D.; Wang, C.; Wutzke, C.; Lemire, G. Biomechanical analysis of maximal instep kick by female soccer players. *J. Hum. Mov. Stud.* **2005**, *49*, 149–168.

54. Shan, G.; Westerhoff, P. Full-body kinematic characteristics of the maximal instep soccer kick by male soccer players and parameters related to kick quality. *Sports Biomech.* **2005**, *4*, 59–72. [CrossRef]

55. Shan, G.; Yuan, J.; Hao, W.; Gu, M.; Zhang, X. Regression equations for estimating the quality of maximal instep kick by males and females in soccer. *Kinesiology* **2012**, *44*, 139–147.

56. Shan, G.; Bohn, C. Anthropometrical data and coefficients of regression related to gender and race. *Appl. Ergon.* **2003**, *34*, 327–337. [CrossRef]

57. Aragón, L.F. Evaluation of four vertical jump tests: Methodology, reliability, validity, and accuracy. *Meas. Phys. Educ. Exerc. Sci.* **2000**, *4*, 215–228. [CrossRef]

58. Shergold, A. The Height of Perfection: Why Cristiano Ronaldo Can Jump Higher than Anyone Else. 2013. Available online: https://www.dailymail.co.uk/sport/football/article-2278671/Cristiano-Ronaldo-Why-Real-Madrid-player-jump-higher-else.html (accessed on 16 July 2018).

59. Blanksby, B.; Nicholson, L.; Elliott, B. Swimming: Biomechanical analysis of the grab, track and handle swimming starts: An intervention study. *Sports Biomech.* **2002**, *1*, 11–24. [CrossRef]

60. UEFA. CRISTIANO RONALDO Great MAN. UNITED #UCL GOALS! 2021. Available online: https://www.youtube.com/watch?v=Z7prW6qpflQ (accessed on 11 November 2021).

61. FIFATV. Zlatan Ibrahimović GOAL—FIFA Puskas Award 2013 WINNER. 2013. Available online: https://www.youtube.com/watch?v=RDsxKXa0PbA (accessed on 11 December 2013).
62. FIFATV. 2014_FIFA Puskas Award—All Nominees Goals. 2014. Available online: https://www.youtube.com/user/FIFATV/videos (accessed on 16 January 2018).

Article

Effects of Shoe Midfoot Bending Stiffness on Multi-Segment Foot Kinematics and Ground Reaction Force during Heel-Toe Running

Ruiya Ma [1], Wing-Kai Lam [2], Rui Ding [3], Fan Yang [3,4] and Feng Qu [1,*]

[1] Biomechanics Laboratory, School of Sport Science, Beijing Sport University, Beijing 100084, China
[2] Sports Information and External Affairs Center, Hong Kong Sports Institute, Hong Kong 999077, China
[3] Li Ning Sports Science Research Center, Li Ning (China) Sports Goods Company Limited, Beijing 101111, China
[4] School of Sports Science, Lingnan Normal University, Zhanjiang 524048, China
* Correspondence: qufeng929@163.com; Tel.: +86-010-6298-9583

Abstract: We investigated how midfoot stiffness of running shoes influences foot segment kinematics and ground reaction force (GRF) during heel-toe running. Nineteen male rearfoot strike runners performed overground heel-toe running at 3.3 m/s when wearing shoes with different midfoot bending stiffnesses (low, medium, and high) in a randomized order. A synchronized motion capture system (200 Hz) and force plate (1000 Hz) were used to collect the foot-marker trajectories and GRF data. Foot kinematics, including rearfoot-lab, midfoot-rearfoot, forefoot-rearfoot, and forefoot-midfoot interactions, and kinetics, including GRF characteristics, were analyzed. Our results indicated that high midfoot stiffness shoes reduced the forefoot-rearfoot range of motion (mean $\pm$ SD; high stiffness, 7.8 $\pm$ 2.0°, low stiffness, 8.7 $\pm$ 2.1°; $p < 0.05$) and forefoot-midfoot range of motion (mean $\pm$ SD; high stiffness, 4.2 $\pm$ 1.1°, medium stiffness, 4.6 $\pm$ 0.9°; $p < 0.05$) in the frontal plane. No differences were found in the GRF characteristics among the shoe conditions. These findings suggest that an increase in midsole stiffness only in the midfoot region can reduce intersegmental foot medial-lateral movements during the stance phase of running. This may further decrease the tension of the foot muscles and tendons during prolonged exercises.

Keywords: midsole modification; overground running; foot kinematics; running injuries; footwear

Citation: Ma, R.; Lam, W.-K.; Ding, R.; Yang, F.; Qu, F. Effects of Shoe Midfoot Bending Stiffness on Multi-Segment Foot Kinematics and Ground Reaction Force during Heel-Toe Running. *Bioengineering* **2022**, *9*, 520. https://doi.org/10.3390/bioengineering9100520

Academic Editors: Rui Zhang and Wei-Hsun Tai

Received: 30 August 2022
Accepted: 28 September 2022
Published: 2 October 2022

Publisher's Note: MDPI stays neutral with regard to jurisdictional claims in published maps and institutional affiliations.

1. Introduction

Running is one of the most popular activities, and footwear is commonly used during running exercises, except for barefoot running. Footwear constructions and features such as midsole hardness, midsole thickness, heel-toe drop, heel counter, longitudinal bending stiffness, and torsional stiffness would significantly alter foot position and joint loading, which is aimed at improving running performance and preventing running injuries [1,2]. Footwear scientists have explored the effect of isolated running shoe features that have the potential to optimize shoe characteristics and performance for decades.

The longitudinal bending stiffness is described as shoe rotation around the medial-lateral axis and is considered to be one of the key features in running shoe development. This can be optimized by incorporating higher-density/thicker midsoles, carbon fiber/thermoplastic polyurethane (TPU) plates, or flex grooves [1]. The foot is frequently considered a rigid segment in running biomechanical research [3–6]. However, intersegmental motion cannot be neglected because the degree of motion is comparable to ankle motion during walking [7–9]. Studying shoe bending stiffness would influence intersegmental foot kinematics and can assist in understanding the intrinsic structural function of the foot during movements.

Previous studies examined the effect of shoe bending stiffness on arch movement. Cigoja et al. [10] suggested that the absolute peak change in the arch angle was about an average of 1.6° smaller in the shoes with a carbon plate than in the control shoe condition. Renan et al. [11] found that shoes with drilled holes in the forefoot increased the forefoot-rearfoot (forefoot with respect to the rearfoot) maximum dorsiflexion angle and the sagittal range of motion during walking. Both studies indicated that increased shoe midsole stiffness is related to decreased arch deformation.

Moreover, Renan et al. [11] found that shoes with lower forefoot stiffness would lead to an increased forefoot-rearfoot maximum inversion angle during walking. The forefoot-rearfoot movement in the frontal plane is described as a foot torsional motion [12], which can be restricted when the shoe torsional stiffness increases [13,14]. Arndt et al. [15] demonstrated that shoe bending and torsional stiffness decreased when a deep groove was cut underneath the midfoot region of the shoe sole. However, no significant changes were found in the intersegmental foot kinematics between the cut and uncut shoes. Since the forefoot-rearfoot motion was reported to be larger during running than in walking [16], the modification of shoe midfoot stiffness might impact running to a greater extent than walking.

Stiff shoes with carbon fiber plates have been reported to systematically reduce the metatarsophalangeal (MTP) maximum dorsiflexion and plantar flexion velocities [17] and peak arch flexion velocity [10] during the propulsion phase of running, which indicates that increased stiffness would reduce foot propulsion velocity. The midfoot stiffness may produce similar results, as the midfoot connects with the rearfoot and forefoot and contributes to the foot transition from heel contact to toe-off.

Shoe bending stiffness also influences the ground reaction force (GRF) data. Shoes with carbon fiber insoles decreased the peak GRF in the anterior-posterior direction. They increased the stance duration while the anterior-posterior GRF impulse and peak vertical GRF remained unchanged [18]. Flores et al. [19] demonstrated that an increased shoe bending stiffness decreases the GRF after the vertical and anterior propulsion peaks during running using statistical parametric mapping. It is unknown whether increased midfoot bending stiffness could decrease the GRF during the stance phase of running.

The effect and underlying mechanism of shoe midfoot stiffness on intersegmental foot movement and GRF during running are still questionable. Hence, the objective of this study was to examine the effect of shoe midfoot bending stiffness on intersegmental foot kinematics using a multi-segment foot model and GRF characteristics. By gradually increasing midfoot stiffness, we hypothesized that stiffer midsoles would exhibit smaller intersegmental motion in the sagittal and frontal planes, slower propulsion velocity, and smaller GRF during running.

2. Materials and Methods

2.1. Participants

A sample size of 12 was calculated based on an analysis of variance (ANOVA: repeated measures, within factors) to provide sufficient statistical power (0.8) to detect a large effect size ($\eta_p^2 = 0.14$, effect size f = 0.403) to examine the differences between the three tested conditions at 0.05 when the correlation among repeated measures was set at 0.5 using G*Power version 3.1.9.7 [20]. Nineteen male recreational runners (aged 22.3 ± 2.5 years, height 1.80 ± 0.04 m, mass 71.3 ± 5.8 kg) were recruited from a local sports university. To fit into the experimental shoes, the participants' heel-toe length was between US 8.5 and 9.5 (The Brannock Device Company, Liverpool, NY, USA). Participants reported in a questionnaire that their average running experience and exposure were 7.6 ± 2.2 years and 15.7 ± 5.6 km/week, respectively. All participants were lower extremity injury-free for at least six months before the study, and they were right-leg dominant, which was determined by the leg they kicked a ball [21]. During data collection, any participant who showed a forefoot or midfoot running strike pattern characterized by the heel marker trajectory reaching its lowest point later than the forefoot/midfoot markers or only a simple peak found in the vertical GRF curve was excluded [22,23]. Each participant signed an informed consent form before the start of data collection. Ethics approval was granted by the Institutional Review Board of Beijing Sport University (IRB Number: 2021146H).

2.2. Footwear Conditions

Three shoe conditions with varying midfoot bending stiffness (high, medium, and low) were built from a neutral running shoe model with a midsole ethylene-vinyl acetate copolymer (EVA) hardness of Asker C 45 (LN ARJH001, Li Ning, Beijing, China). The medium stiffness shoe was identical to the original shoe model. The low stiffness shoe was modified by cutting with three 1 cm depth and 0.9–1.8 cm width transverse grooves at the midfoot region. In contrast, the high stiffness shoe was modified by replacing the original EVA midfoot plate with a stiffer TPU midfoot plate (Shore D 70). All three shoe conditions had the same size (9.5 cm × 2.5 cm × 0.2 cm) and location of the midfoot shank plate as per the manufacturer's specification (Figure 1A). The longitudinal midfoot bending stiffness was quantified using a mechanical shoe flexion tester (Exeter Research, Brentwood, NH, USA) by bending across the midfoot flexion axis located 45% of the shoe length from the heel for each shoe (Figure 1B). Sixty consecutive flexion-extension cycles were performed and the stiffness between 10° and 25° of the 51st to 55th trials was averaged to denote the midfoot bending stiffness (modified according to the forefoot flexibility test standard: ASTM 911-85). The bending stiffness values for the low, medium, and high stiffness shoes were 0.069, 0.091, and 0.163 Nm/°, respectively.

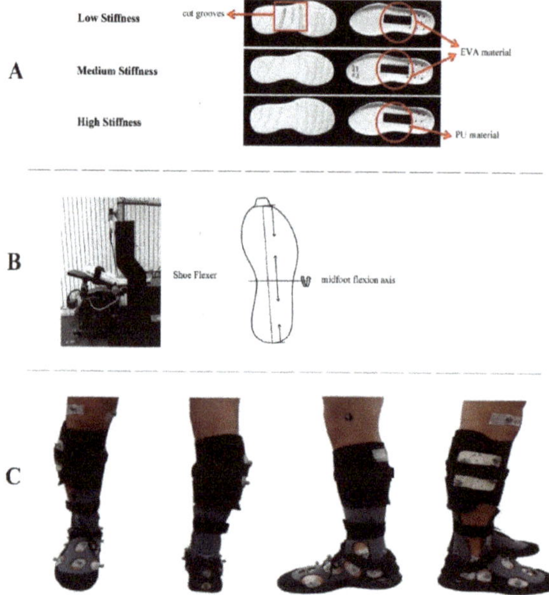

Figure 1. (**A**) Constructions of shoe conditions (low, medium, and high), with ethylene-vinyl acetate copolymer (EVA) or thermoplastic polyurethane (TPU) plate in the midfoot, (**B**) Bending axis of midfoot mechanical flexion test, and (**C**) Marker placement on the body with the subject shod.

There are different approaches to marker attachment for foot kinematics measurement: shoe surface markers, foot skin markers, and bone-mounted markers. Arnold and Bishop [24] reviewed articles and concluded that it is inappropriate to describe in-shoe foot motion with markers placed on the external shoe surface owing to the induced errors and lack of validity. However, the bone-mounted marker approach is not widely accepted because of its invasive preparation procedures [15]. Attaching markers directly to the skin by drilling holes in the shoes is an alternative approach for studying multi-segment foot motion [25,26]. Furthermore, to accurately assess the actual foot motion, Sterzing et al. [27] removed the original shoe upper and modified the upper with custom-made sock technology, which was thin and made of neoprene material for stretch boots. In the current study, we removed the original shoe uppers and maintained the heel counter and a small part of the toe counter. Neoprene shoe socks (Ultra Stretch Boots, Cressi-Sub, Genoa, Italy) were then glued securely to the sole of each tested shoe (high, medium, and low stiffness). Elliptical holes were prepared at anatomical landmarks to allow direct skin placement of markers with diameters between 25 and 40 mm [28]. All shoe soles of the experimental shoes were painted black to minimize the potential visual impact across the footwear conditions (Figure 1C). Participants were not informed of the actual differences between the shoes.

2.3. Procedures

All reflective markers were placed at marker locations on the skin by the same experienced researcher across all tested conditions and participants (Figures 1C and 2). Eight markers at the lateral and medial malleoli (LM and MM), lateral and medial epicondyle of the femur (LE and ME), and shank board (SK1-4) were added to our marker set to build the shank and foot coordinate systems and track shank movement. The first metatarsal head and base (MH1 and MB1), second metatarsal head and base (MH2 and MB2), fifth metatarsal head and base (MH5 and MB5), navicular tuberosity (NV), calcaneus posterior surface (PC), the lateral apex of the peroneal tubercle (LC), and most medial apex of the sustentaculum tali (MC) were attached over the right foot to define the forefoot, midfoot, and rearfoot segments based on the recommendations suggested by Leardini et al. [29].

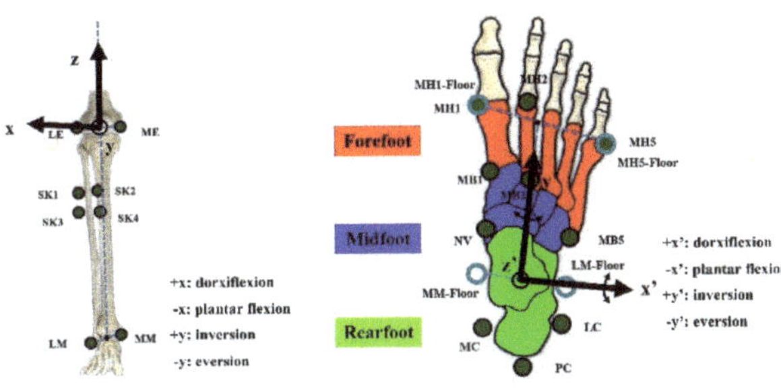

Figure 2. (**A**). A shank coordinate system (xyz) was built to calculate rearfoot-lab and rearfoot-shank kinematics. (**B**). A foot coordinate system (x'y'z') was built to quantify foot complex motion. Illustration of the marker locations to define the rearfoot (green area), midfoot (dark blue area), and forefoot (red area). Markers at the lateral and medial malleoli (LM and MM), lateral and medial epicondyle of the femur (LE and ME), shank board (SK1-4), first metatarsal head and base (MH1 and MB1), second metatarsal head and base (MH2 and MB2), fifth metatarsal head and base (MH5 and MB5), navicular tuberosity (NV), calcaneus posterior surface (PC), the lateral apex of the peroneal tubercle (LC), and most medial apex of the sustentaculum tali (MC) were attached to the subject.

According to previous studies [27,30], the bases of the markers were securely attached to their locations throughout the experiment. For each participant, only the balls (spherical markers) were removed from the shoe condition already measured and placed back onto the same bases on the next shoe condition to allow consistent marker locations across the footwear conditions. This arrangement ensured identical marker locations throughout the running trials under all three footwear conditions. The participants' feet were disinfected before each test, and the neoprene socks were blow-dried after each use to avoid bacterial infection.

A six-camera motion capture system (200 Hz; Vicon, Metrics Ltd., Oxford, UK) was synchronized with a force plate (1000 Hz; OR6GT, 90 × 60 cm, AMTI, Watertown, MA, USA) to collect lower extremity kinematics and GRFs during running. Running trials were performed over a 16 m laboratory runway with a force plate mounted 8 m from the start line. The target running speed was set at $3.3 \pm 5\%$ m/s [31] and was controlled by two timing gates (Brower Timing Systems, Salt Lake City, UT, USA), which were placed 3.3 m apart in the middle of the runway. Dominant leg heel strike trials at the target speed with all tracking markers in place and where the entire right foot contacted within the force plate surface during the stance phase were considered successful trials. Practice trials were conducted to familiarize the participants with the test procedure and running speed before data collection. Each participant performed five successful trials for each condition. The order of the shoe conditions was randomized across the participants.

2.4. Data Processing

Fourth-order Butterworth low-pass filters with cut-off frequencies of 150 and 30 Hz were applied to the GRF and kinematic data [31,32], respectively. The stance phase was defined from heel contact to toe-off and determined using vertical GRF data at a threshold of 10 N [33]. The entire stance phase was divided into braking and propulsion phases when the time of the anterior-posterior GRF data crossed zero [33–35].

A shank coordinate system (xyz) was built to calculate the segmental rotation (rearfoot-lab, rearfoot segment with respect to the laboratory coordinate frame) and joint rotation (rearfoot-shank, rearfoot segment with respect to the shank) kinematics (Figure 2A). The origin was set at the midpoint of the ME and LE, the z-axis pointed upward and along the line connecting the origin with the midpoint of MM and LM, and the x-axis was orthogonal to an anatomical plane, which was defined among four markers (LM, MM, ME, and LE) using the least squares fit, and the y-axis was orthogonal to the x–z plane and followed the right-hand rule. A foot coordinate reference frame $(x'y'z')$ was built in our study to calculate joint angles, including midfoot-rearfoot (midfoot with respect to the rearfoot), forefoot-rearfoot (forefoot with respect to the rearfoot, a non-adjacent joint) and forefoot-midfoot (forefoot with respect to the midfoot) (Figure 2B). The origin of the reference frame was defined as the midpoint of the two landmarks, which were projected onto the ground plane by MM and ML. The direction of the anterior-posterior (y') axis joined the origin with the midpoint of two ground-projected landmarks by MH1 and MH5, and the vertical (z') axis was upward and orthogonal to the transverse plane. The x'-axis was orthogonal to the y'–z' plane, followed by the right-hand rule. All the foot segments share the same foot coordinate system, which allows segments with the same axes of rotation and translation as those of the foot. Specific segmental anatomical tracking markers were defined for each foot segment [29], and the shank was tracked with SK1-4 (Table 1).

Table 1. Tracking markers for foot and shank segments.

Segment	Tracking Markers
Rearfoot	PC, LC, and MC
Midfoot	MB2, MB5, and NV
Forefoot	MH1, MH2, MH5, MB1, and MB5
Shank	SK1-4

Segmental and joint rotations were calculated using Euler angles (sequence sagittal, frontal, and transverse plane motion) according to the International Society of Biomechanics recommendations [36,37], that is, dorsiflexion/plantar flexion as the rotation about the x-axis in the sagittal plane, eversion/inversion about the y-axis in the frontal plane, and internal/external rotation about the z-axis in the transverse plane. During the stance phase, the rearfoot, midfoot, and forefoot kinematics were investigated in the sagittal and frontal planes. Each participant performed a static calibration trial at the neutral standing position before dynamic running trials for each shoe condition, joint angles were defined to be 0° at this position. Dynamic joint angles were calculated relevant to the static calibration trial during the dynamic test. Kinematic data were normalized to 0–100% of the stance phase.

The vertical GRF variables included ground contact time, peak vertical force 1, maximum vertical loading rate 1, peak vertical force 2, and maximum vertical loading rate 2 (Figure 3). Ground contact time was defined as the time of the stance phase, which was calculated as the time between heel contact and toe-off. Vertical loading rate 1 was defined as the maximum instantaneous slope of the vertical GRF time profile between the time of heel contact and the time of peak vertical force 1, and vertical loading rate 2 was computed as the maximum instantaneous slope of the vertical GRF time profile between the time of peak vertical force 1 and the time of peak vertical force 2. The extracted anterior-posterior GRF variables were peak braking force, peak propulsion force, braking impulse, and propulsion impulse. The braking and propulsion impulses were calculated as the force-time integral during the braking and propulsion phases, respectively. All force-related variables were normalized to body weight (BW). The kinematic and GRF data from the five trials were averaged for each participant and condition.

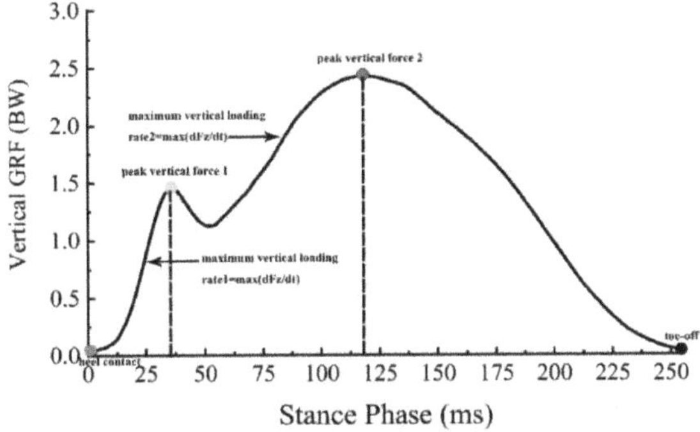

Figure 3. The time profile of vertical ground reaction force (GRF) during the stance phase for one representative trial. The peak vertical force 1 and 2 were shown in blue and red dots, respectively.

2.5. Statistical Analysis

Kolmogorov–Smirnov tests were applied to verify that all data were normally distributed. One-way repeated-measures ANOVAs were performed to detect significant differences in shoe midfoot stiffness on the analyzed parameters. When a main effect of the shoe was present, post hoc tests with Bonferroni correction were applied to detect significant differences between every two conditions. Greenhouse–Geisser's was chosen as an adjustment of the p-value when Mauchly's test of sphericity assumption was rejected. η_p^2 was calculated to show the effect size of ANOVA and was categorized as small ($0.01 \leq \eta_p^2 < 0.06$), medium ($0.06 \leq \eta_p^2 < 0.14$), and large ($\eta_p^2 \geq 0.14$) [38]. Statistical significance was set at $p < 0.05$. All statistical analyses were performed using SPSS 26.0 (IBM Corporation, Chicago, IL, USA).

3. Results

3.1. Kinematics

The multi-segment foot kinematics, including rearfoot-lab, rearfoot-shank, midfoot-rearfoot, forefoot-rearfoot, and forefoot-midfoot interactions of the shoe conditions are displayed in Figure 4. Angle curves in the sagittal and frontal planes and angular velocity curves in the sagittal plane were averaged for all participants and normalized to 0–100% during the stance phase. The 0° angle outlined in the sagittal and frontal planes indicated that the angle of the joint was consistent with the static calibration trial. The rearfoot-lab, rearfoot-shank, midfoot-rearfoot, forefoot-rearfoot, and forefoot-midfoot kinematic variables are provided in Table 2.

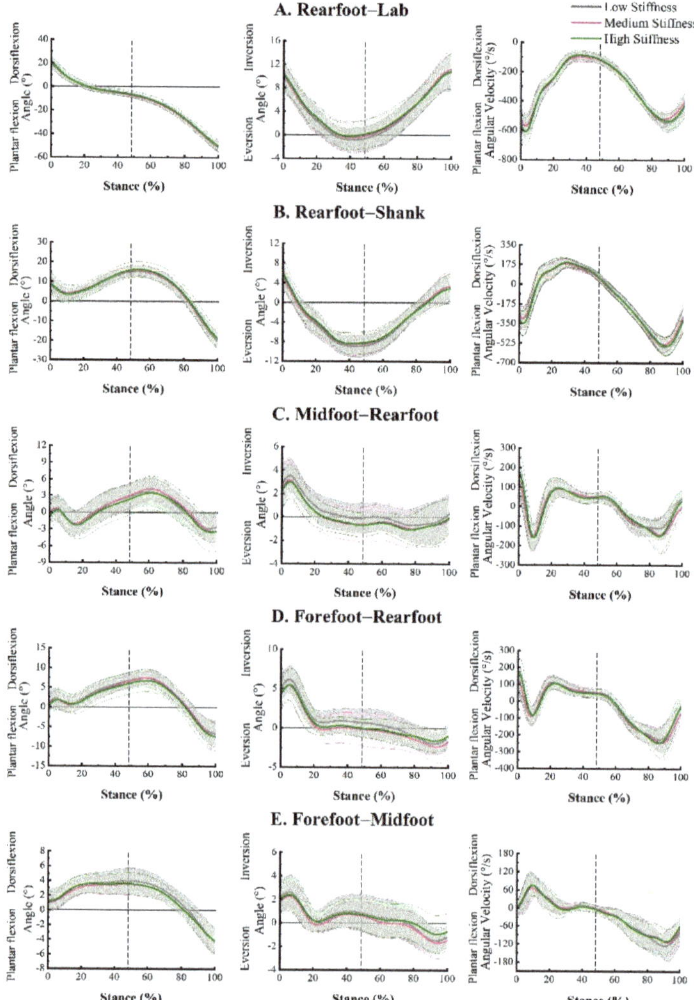

Figure 4. Averaged angle-time curves displayed in the sagittal and frontal planes (first and second columns, respectively) and angular velocity-time curves in the sagittal plane (third column) for rearfoot-lab (**A**), rearfoot-shank (**B**), midfoot-rearfoot (**C**), forefoot-rearfoot (**D**), and forefoot-midfoot (**E**) movements during running stance phase while wearing three shoe conditions. Grey, pink, and green curves separately indicate the low stiffness, medium stiffness, and high stiffness conditions. The dashed line reflects the anterior-posterior force crossed zero.

For rearfoot-lab interactions in the sagittal plane, the dorsiflexion angle at initial contact ($F_{(2, 36)} = 3.270$, $p = 0.050$, $\eta_p^2 = 0.154$) did not show any significant difference. A significant main effect of the shoe was observed for peak plantar flexion angular velocity during the braking phase ($F_{(2, 36)} = 3.697$, $p = 0.035$, $\eta_p^2 = 0.170$). However, Bonferroni post hoc comparisons did not find any difference between shoe conditions. No significant difference was found in rearfoot-shank peak eversion angle and frontal plane range of motion between shoe conditions ($p > 0.05$).

Table 2. Multi-segment kinematic variables (Mean $\pm$ SD [95% confidence interval]) of experimental shoes. A significant main effect *p*-value is bolded.

Segment Interaction	Variable	Low Stiffness	Medium Stiffness	High Stiffness	*p*-Value	η_p^2
Rearfoot-lab	Dorsiflexion angle at initial contact (°)	19.6 ± 4.1 [17.6, 21.6]	20.5 ± 4.9 [18.1, 22.8]	21.1 ± 4.8 [18.8, 23.4]	0.050	0.154
	Peak plantar flexion angular velocity during the braking phase (°/s)	577.4 ± 73.5 [542.0, 612.8]	580.5 ± 81.5 [541.2, 619.8]	617.4 ± 96.0 [571.1, 663.6]	**0.035**	0.170
Rearfoot-shank	Peak eversion angle (°)	9.1 ± 2.2 [8.0, 10.1]	8.7 ± 2.2 [7.6, 9.8]	−8.5 ± 2.2 [7.5, 9.6]	0.360	0.055
	Frontal plane range of motion (°)	14.2 ± 3.1 [12.7, 15.6]	14.0 ± 4.0 [12.0, 15.9]	14.5 ± 3.9 [12.6, 16.4]	0.389	0.051
Midfoot-rearfoot	Peak dorsiflexion angle (°)	4.1 ± 2.4 [2.9, 5.2]	4.5 ± 2.1 [3.5, 5.5]	3.9 ± 2.8 [2.6, 5.3]	0.630	0.025
	Frontal plane range of motion (°)	5.3 ± 2.0 [4.3, 6.2]	5.2 ± 1.9 [4.3, 6.0]	5.1 ± 2.0 [4.2, 6.1]	0.812	0.012
	Peak plantar flexion angular velocity during the braking phase (°/s)	188.5 ± 37.7 [170.3, 206.6]	176.2 ± 64.4 [145.1, 207.2]	182.1 ± 51.5 [157.2, 206.9]	0.653	0.023
	Peak plantar flexion angular velocity during the propulsion phase (°/s)	151.7 ± 41.2 [131.8, 171.5]	177.3 ± 62.7 [147.1, 207.5]	172.1 ± 70.2 [138.2, 205.9]	0.080	0.143
Forefoot-rearfoot	Peak dorsiflexion angle (°)	7.7 ± 2.3 [6.6, 8.8]	7.7 ± 1.6 [7.0, 8.5]	7.0 ± 2.7 [5.7, 8.3]	0.428	0.046
	Frontal plane range of motion (°)	8.7 ± 2.1 [7.7, 9.7]	8.4 ± 2.0 [7.4, 9.4]	7.8 ± 2.0 [6.8, 8.8]	**0.005** [a]	0.258
	Peak plantar flexion angular velocity during the braking phase (°/s)	109.8 ± 40.3 [90.4, 129.3]	106.2 ± 46.2 [83.9, 128.5]	100.8 ± 47.5 [77.9, 123.7]	0.680	0.021
	Peak plantar flexion angular velocity during the propulsion phase (°/s)	244.9 ± 55.7 [218.0, 271.7]	266.1 ± 77.6 [228.7, 303.6]	265.5 ± 63.3 [235.0, 296.0]	0.066	0.140
Forefoot-midfoot	Peak dorsiflexion angle (°)	4.4 ± 1.7 [3.5, 5.2]	3.9 ± 1.4 [3.3, 4.6]	4.3 ± 1.3 [3.6, 4.9]	0.135	0.105
	Frontal plane range of motion (°)	4.7 ± 1.3 [4.1, 5.4]	4.6 ± 0.9 [4.2, 5.0]	4.2 ± 1.1 [3.7, 4.8]	**0.020** [b]	0.196
	Peak dorsiflexion angular velocity during the braking phase (°/s)	102.0 ± 33.5 [85.9, 118.2]	95.2 ± 32.0 [79.8, 110.6]	102.5 ± 34.0 [86.1, 118.9]	0.512	0.036
	Peak plantar flexion angular velocity during the propulsion phase (°/s)	154.4 ± 46.0 [132.2, 176.6]	144.8 ± 35.8 [127.6, 162.0]	136.5 ± 32.7 [120.7, 152.2]	**0.043**	0.182

[a] significant difference in low stiffness and high stiffness conditions. [b] significant difference in medium stiffness and high stiffness conditions.

None of the midfoot-rearfoot variables, including the peak dorsiflexion angle, frontal plane range of motion, and peak plantar flexion angular velocities, showed any differences in sagittal and frontal plane movements between shoes during braking and propulsion phases ($p > 0.05$).

A significant main effect (F (2, 36) = 6.247, $p = 0.005$, $\eta_p^2 = 0.258$) was observed in the frontal range of motion of the forefoot-rearfoot joint. The magnitude was greater in the low stiffness shoe (8.7 ± 2.1°) than that in the high stiffness shoe (7.8 ± 2.0°) during the stance phase.

The forefoot-midfoot variable indicated a significantly greater frontal plane range of motion (F (2, 36) = 4.399, $p = 0.020$, $\eta_p^2 = 0.196$) in the medium stiffness shoe (4.6 ± 0.9°) than in the high stiffness shoe (4.2 ± 1.1°). A significant main effect (F (1.430, 25.747) = 4.009, $p = 0.043$, $\eta_p^2 = 0.182$) of the shoe condition was found when comparing the peak plantar flexion velocity during the propulsion phase, and no significant difference was observed for post hoc comparisons.

3.2. GRF

The averaged vertical and anterior-posterior GRF curves are displayed in Figure 5, which show very similar patterns across shoe conditions during the stance phase of running. No significant differences were found in the ground contact time, peak vertical force 1, maximum vertical loading rate 1, peak vertical force 2, maximum vertical loading rate 2, peak braking force, peak propulsion force, braking impulse, and propulsion impulse between shoe conditions ($p > 0.05$; Table 3).

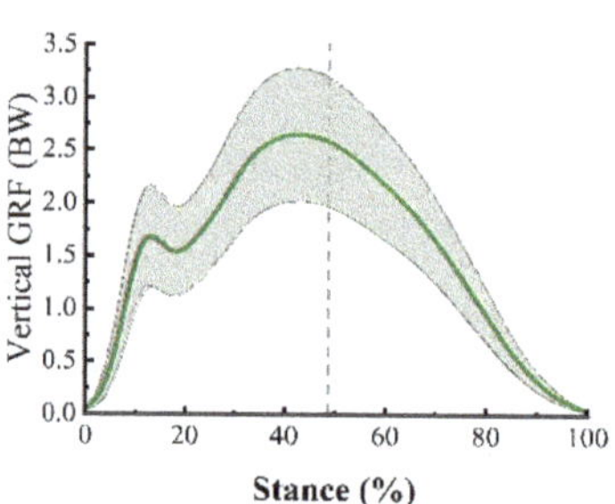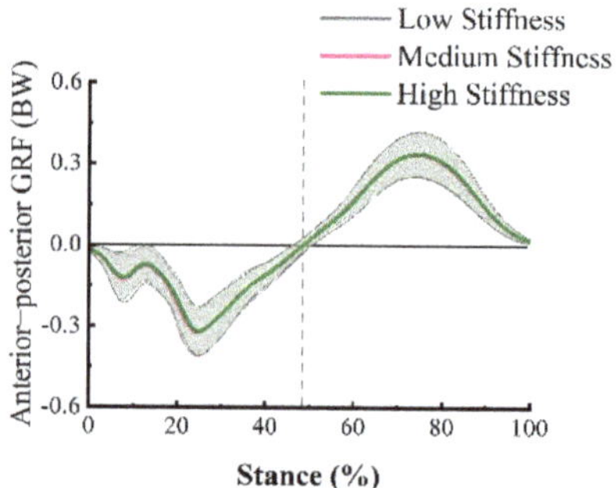

Figure 5. Averaged force-time curves displayed the vertical and anterior-posterior GRF. Grey, pink, and green curves separately indicate the low stiffness, medium stiffness, and high stiffness conditions. The dashed line reflects the anterior-posterior force crossed zero.

Table 3. GRF-related variables (Mean $\pm$ SD [95% confidence interval]) of experimental shoes.

Variable	Low Stiffness	Medium Stiffness	High Stiffness	*p*-Value	η_p^2
Ground contact time (ms)	232.8 $\pm$ 17.3 [224.5, 241.1]	233.4 $\pm$ 15.9 [225.8, 241.1]	233.4 $\pm$ 16.2 [225.6, 241.2]	0.870	0.008
Peak vertical force 1 (BW)	1.74 $\pm$ 0.27 [1.61, 1.87]	1.75 $\pm$ 0.30 [1.61, 1.90]	1.75 $\pm$ 0.31 [1.60, 1.90]	0.858	0.008
Maximum vertical loading rate 1 (BW/s)	105.9 $\pm$ 25.2 [93.8, 118.1]	109.9 $\pm$ 29.5 [95.7, 124.1]	107.8 $\pm$ 27.61 [94.5, 121.1]	0.941	0.003
Peak vertical force 2 (BW)	2.67 $\pm$ 0.22 [2.56, 2.77]	2.67 $\pm$ 0.21 [2.60, 2.77]	2.67 $\pm$ 0.22 [2.57, 2.78]	0.336	0.059
Maximum vertical loading rate 2 (BW/s)	39.58 $\pm$ 6.69 [36.36, 42.81]	39.00 $\pm$ 5.84 [36.19, 41.82]	39.37 $\pm$ 6.96 [36.01, 42.72]	0.743	0.016
Peak braking force (BW)	0.34 $\pm$ 0.01 [0.33, 0.36]	0.34 $\pm$ 0.01 [0.33, 0.36]	0.35 $\pm$ 0.01 [0.33, 0.36]	0.886	0.008
Peak propulsion force (BW)	0.35 $\pm$ 0.01 [0.32, 0.38]	0.35 $\pm$ 0.01 [0.32, 0.37]	0.34 $\pm$ 0.01 [0.31, 0.37]	0.340	0.058
Braking impulse (Ns/BW)	0.016 $\pm$ 0.003 [0.018, 0.014]	0.016 $\pm$ 0.003 [0.018, 0.015]	0.016 $\pm$ 0.004 [0.018, 0.014]	0.574	0.030
Propulsion impulse (Ns/BW)	0.023 $\pm$ 0.003 [0.021, 0.024]	0.023 $\pm$ 0.003 [0.021, 0.024]	0.023 $\pm$ 0.003 [0.021, 0.024]	0.886	0.007

4. Discussion

This study examined the influence of shoe midfoot stiffness on multi-segment foot movements and GRF during heel-toe running. The results of this study provide insights into the underlying mechanism of footwear midfoot stiffness on foot joint kinematics and GRF characteristics. Our study showed similar patterns of foot segmental motion across shoe conditions during the stance phase of running (Figure 4). However, the results of this study indicate that changes in midfoot stiffness can influence foot motion between high stiffness shoes and low stiffness or medium stiffness shoes. Our present study found no significant differences between low stiffness and medium stiffness conditions for the intersegmental foot kinematics in both the sagittal and frontal planes, which may be due to the small mechanical property differences between shoes.

Our study's initial rearfoot-lab dorsiflexion angle >8.0° confirmed that all participants performed rearfoot strike patterns [39]. The influence of shoe midfoot stiffness on the initial rearfoot-lab dorsiflexion angle did not show any difference. This is in line with a previous study on shoe forefoot hardness, which showed the same sagittal shoe-ground angle at the initial ground contact [31]. These results may be explained by the fact that the rearfoot structure and material were identical for the low, medium, and high midfoot stiffness shoe conditions, resulting in similar ground contact kinematics.

During the propulsion phase, the participants displayed peak midfoot-rearfoot, forefoot-rearfoot, and forefoot-midfoot plantar flexion velocities. It is assumed that the increased midfoot bending stiffness could increase the resistance in the midfoot region, which could decrease the plantar flexion speed during the propulsion phase. Previous studies also found that an increase in shoe stiffness decreases the maximum MTP plantar flexion [17] and peak arch flexion velocities [10] during the terminal stance. Our results indicated that midfoot bending stiffness had a significant main effect on forefoot-midfoot peak plantar flexion angular velocity during the propulsion phase. Still, no significant difference could be determined between shoe conditions. Larger effects may be possible if greater modifications are made to the experimental shoes.

Increased shoe forefoot [11] and full-length [10] midsole bending stiffness reduced arch deformation during walking and running, respectively. However, none of the midfoot-rearfoot, forefoot-rearfoot, and forefoot-midfoot peak dorsiflexion angles showed any significant difference between shoe conditions in our study. The results of the current study suggest that alterations in midfoot stiffness would not be sufficient to vary the amount of midfoot deformation. This may be explained by the fact that the midfoot plate was too short to generate a lever arm effect on arch movement.

Participants wearing shoes with higher midfoot bending stiffness demonstrated smaller forefoot-midfoot and forefoot–rearfoot frontal plane range of motion during the running stance phase. Still, no significant differences were found for the midfoot-rearfoot joint. Renan et al. [11] found that shoes with drilled holes in the forefoot increased the forefoot-rearfoot inversion angle in the frontal plane. However, Arndt et al. [15] showed no increase in frontal plane ranges of motion when walking with shoes with lower midfoot bending stiffness. The differences across studies may be due to the small number of participants or insufficient cuts to create a softer stiffness prototype in their study and the higher exercise intensity of the participants in our study. The forefoot-rearfoot motion in the frontal plane is considered a foot torsional motion which can be decreased when the shoe torsional stiffness increases [13,14]. Although we did not quantify the torsional stiffness of the shoe conditions, the decreased torsional movement may be explained by the increased shoe torsional stiffness under the high bending stiffness condition.

Previous studies indicated that increased foot torsional stiffness restricts the forefoot-rearfoot motion in the frontal plane, and thus leads to an excessive motion of the ankle pronation [12,14], which is considered a risk factor for running-related injuries [40–42]. Stacoff et al. found that running with a shoe decreases the torsion movement of the foot, and increases the pronation of the foot compared with barefoot running [12]. Another study by Graf et al. observed a larger ankle range of motion with increasing shoe torsional stiffness [14], whereas Graf and Stefanyshyn [13] did not find any difference in ankle kinematics for shoes with different torsion ranges of motion. In our study, midfoot stiffness did not affect rearfoot-shank motion in the frontal plane, suggesting that the increase in midfoot stiffness alone may have minimal effect on pronation-related injuries.

No significant differences were found in the peak vertical forces and instantaneous loading rates between the shoe conditions. Oh and Park [18] also found that shoes with carbon fiber insoles did not influence vertical GRF data, which indicates that the change in shoe bending stiffness at the midfoot area would not influence the vertical GRF. Moreover, Oh and Park [18] investigated a longer stance duration and a smaller anterior-posterior GRF when wearing shoes with carbon fiber insoles, but no significant difference was found for the anterior-posterior GRF impulse. In our study, the ground contact time, peak anterior-posterior forces, and the anterior-posterior impulses did not differ among shoe conditions. The participants maintained their anterior-posterior impulses to keep the same running velocity across the different shoes.

Some limitations of this study must be acknowledged when interpreting the results. First, the shoe upper was mostly modified, which may have sacrificed the original function of the shoe support (e.g., shoe upper stability). However, the elastic sock upper is a good alternative to bone-pin markers for the non-invasive measurement of actual foot multi-segment. Second, only laboratory ground running was measured in the present study and studying midfoot stiffness during long-distance running should be considered. Third, although our primary scope of study aimed to analyze the actual multi-segment foot kinematics, running shoe midfoot stiffness might influence the other proximal joint kinematics (ankle, knee, or hip) according to the kinetic chain theory. Fourth, this study did not analyze the multi-segment foot kinetic influences of running shoe midfoot stiffness as no reliable anthropometric data are available, such as the center of mass, for accurate foot segment calculation. Fifth, only males were recruited in our study as the tested shoes were built in US size 9, and the current results may not be generalized to females. Finally, we just tested three shoe conditions whose stiffness are 0.069, 0.091, and 0.163 Nm/°, and future research should consider a larger range of midfoot stiffness conditions to optimize the midfoot stiffness in the running community.

5. Conclusions

The midfoot stiffness of running shoes can influence multi-segment foot movements during heel-toe running. Shoes with increased midfoot bending stiffness can reduce foot medial-lateral movements during the stance phase of running, indicating that the length variation of intersegmental muscles and tendons should also be smaller, which may decrease the tension of the foot muscles and tendons connected between the metatarsals and tarsus. However, we found no difference between shoe conditions in ankle kinematics, thus the effect of shoe midfoot stiffness related to pronation injury was not evident.

Author Contributions: Conceptualization, R.M., W.-K.L., R.D. and F.Q.; methodology, R.M., W.-K.L., R.D. and F.Q.; software, R.M. and R.D.; validation, W.-K.L. and F.Y.; formal analysis, W.-K.L. and F.Y.; investigation, R.D. and F.Y.; resources, W.-K.L. and R.D.; data curation, R.M.; writing—original draft preparation, R.M.; writing—review and editing, R.M., W.-K.L., R.D., F.Y. and F.Q.; visualization, R.M.; supervision, F.Q.; project administration, F.Y. and F.Q. All authors have read and agreed to the published version of the manuscript.

Funding: This research received no external funding.

Institutional Review Board Statement: The study was conducted in accordance with the Declaration of Helsinki, and approved by the Institutional Review Board of Beijing Sport University (protocol code: 2021146H and date of approval: 10 October 2021).

Informed Consent Statement: Informed consent was obtained from all subjects involved in the study.

Data Availability Statement: The data presented in this study are available on request from the corresponding author.

Acknowledgments: The authors would like to thank Thorsten Sterzing for the study design and all the advice provided for this research.

Conflicts of Interest: The authors declare no conflict of interest.

References

1. Hoitz, F.; Mohr, M.; Asmussen, M.; Lam, W.K.; Nigg, S.; Nigg, B. The effects of systematically altered footwear features on biomechanics, injury, performance, and preference in runners of different skill level: A systematic review. *Footwear Sci.* **2020**, *12*, 193–215. [CrossRef]
2. Sun, X.; Lam, W.K.; Zhang, X.; Wang, J.; Fu, W. Systematic review of the role of footwear constructions in running biomechanics: Implications for running-related injury and performance. *J. Sports Sci. Med.* **2020**, *19*, 20–37. [PubMed]
3. Hannigan, J.J.; Pollard, C.D. Differences in running biomechanics between a maximal, traditional, and minimal running shoe. *J. Sci. Med. Sport* **2020**, *23*, 15–19. [CrossRef]
4. Hardin, E.C.; van den Bogert, A.J.; Hamill, J. Kinematic adaptations during running: Effects of footwear, surface, and duration. *Med. Sci. Sports Exerc.* **2004**, *36*, 838–844. [CrossRef] [PubMed]
5. Morio, C.; Flores, N. Effect of shoe bending stiffness on lower limb kinetics of female recreational runners. *Comput. Methods Biomech. Biomed. Eng.* **2017**, *20*, 139–140. [CrossRef]
6. Squadrone, R.; Rodano, R.; Hamill, J.; Preatoni, E. Acute effect of different minimalist shoes on foot strike pattern and kinematics in rearfoot strikers during running. *J. Sport Sci.* **2015**, *33*, 1196–1204. [CrossRef]
7. Dubbeldam, R.; Buurke, J.H.; Simons, C.; Groothuis-Oudshoorn, C.G.; Baan, H.; Nene, A.V.; Hermens, H.J. The effects of walking speed on forefoot, hindfoot and ankle joint motion. *Clin. Biomech.* **2010**, *25*, 796–801. [CrossRef]
8. Lundgren, P.; Nester, C.; Liu, A.; Arndt, A.; Jones, R.; Stacoff, A.; Wolf, P.; Lundberg, A. Invasive in vivo measurement of rear-, mid- and forefoot motion during walking. *Gait Posture* **2008**, *28*, 93–100. [CrossRef]
9. Stebbins, J.; Harrington, M.; Thompson, N.; Zavatsky, A.; Theologis, T.J.G. Posture. Repeatability of a model for measuring multi-segment foot kinematics in children. *Gait Posture* **2006**, *23*, 401–410. [CrossRef] [PubMed]
10. Cigoja, S.; Asmussen, M.J.; Firminger, C.R.; Fletcher, J.R.; Edwards, W.B.; Nigg, B.M. The effects of increased midsole bending stiffness of sport shoes on muscle-tendon unit shortening and shortening velocity: A randomised crossover trial in recreational male runners. *Sports Med. Open* **2020**, *6*, 9. [CrossRef] [PubMed]

11. Renan, A.; Resende, T.S.; Fonseca, P.L.; Silva, A.E.; Pertence, R.; Kirkwood, N. Forefoot midsole stiffness affects forefoot and rearfoot kinematics during the stance phase of gait. *J. Am. Podiatr. Med. Assoc.* **2014**, *104*, 183–190.
12. Stacoff, A.; Kaelin, X.; Stuessi, E.; Segesser, B. The torsion of the foot in running. *J. Appl. Biomech.* **1989**, *5*, 375–389. [CrossRef]
13. Graf, E.; Stefanyshyn, D. The effect of shoe torsional stiffness on lower extremity kinematics and biomechanical risk factors for patellofemoral pain syndrome during running. *Footwear Sci.* **2012**, *4*, 199–206. [CrossRef]
14. Graf, E.; Wannop, J.W.; Schlarb, H.; Stefanyshyn, D. Effect of torsional stiffness on biomechanical variables of the lower extremity during running. *Footwear Sci.* **2017**, *9*, 1–8. [CrossRef]
15. Arndt, A.; Lundgren, P.; Liu, A.; Nester, C.; Maiwald, C.; Jones, R.; Lundberg, A.; Wolf, P.J.F.E. The effect of a midfoot cut in the outer sole of a shoe on intrinsic foot kinematics during walking. *Footwear Sci.* **2013**, *5*, 63–69. [CrossRef]
16. Morio, C.; Lake, M.J.; Gueguen, N.; Rao, G.; Baly, L. The influence of footwear on foot motion during walking and running. *J. Biomech.* **2009**, *42*, 2081–2088. [CrossRef]
17. Willwacher, S.; König, M.; Potthast, W.; Brüggemann, G.-P. Does specific footwear facilitate energy storage and return at the metatarsophalangeal joint in running? *J. Appl. Biomech.* **2013**, *29*, 583–592. [CrossRef] [PubMed]
18. Oh, K.; Park, S. The bending stiffness of shoes is beneficial to running energetics if it does not disturb the natural MTP joint flexion. *J. Biomech.* **2017**, *53*, 127. [CrossRef]
19. Flores, N.; Delattre, N.; Berton, E.; Rao, G. Effects of shoe energy return and bending stiffness on running economy and kinetics. *Footwear Sci.* **2017**, *9*, S11–S13. [CrossRef]
20. Faul, F.; Erdfelder, E.; Lang, A.-G.; Buchner, A. G*Power 3: A flexible statistical power analysis program for the social, behavioral, and biomedical sciences. *Behav. Res. Methods* **2007**, *39*, 175–191. [CrossRef] [PubMed]
21. Peters, M. Footedness: Asymmetries in foot preference and skill and neuropsychological assessment of foot movement. *Psychol. Bull.* **1988**, *103*, 179–192. [CrossRef]
22. Altman, A.R.; Davis, I.S. Barefoot running: Biomechanics and implications for running injuries. *Curr. Sport Med. Rep.* **2012**, *11*, 244–250. [CrossRef]
23. Lieberman, D.E.; Venkadesan, M.; Werbel, W.A.; Daoud, A.I.; D'Andrea, S.; Davis, I.S.; Mang'eni, R.O.; Pitsiladis, Y. Foot strike patterns and collision forces in habitually barefoot versus shod runners. *Nature* **2010**, *463*, 531–535. [CrossRef] [PubMed]
24. Arnold, J.B.; Bishop, C.J.F.E. Quantifying foot kinematics inside athletic footwear: A review. *Footwear Sci.* **2013**, *5*, 55–62. [CrossRef]
25. Wegener, C.; Greene, A.; Burns, J.; Hunt, A.E.; Vanwanseele, B.; Smith, R.M. In-shoe multi-segment foot kinematics of children during the propulsive phase of walking and running. *Hum. Mov. Sci.* **2015**, *39*, 200–211. [CrossRef] [PubMed]
26. Wolf, S.; Simon, J.; Patikas, D.; Schuster, W.; Armbrust, P.; Döderlein, L. Foot motion in children shoes—A comparison of barefoot walking with shod walking in conventional and flexible shoes. *Gait Posture* **2008**, *27*, 51–59. [CrossRef]
27. Sterzing, T.; Custoza, G.; Ding, R.; Cheung, T.M. Segmented midsole hardness in the midfoot to forefoot region of running shoes alters subjective perception and biomechanics during heel-toe running revealing potential to enhance footwear. *Footwear Sci.* **2015**, *7*, 63–79. [CrossRef]
28. Bishop, C.; Arnold, J.B.; Fraysse, F.; Thewlis, D. A method to investigate the effect of shoe-hole size on surface marker movement when describing in-shoe joint kinematics using a multi-segment foot model. *Gait Posture* **2015**, *41*, 295–299. [CrossRef] [PubMed]
29. Leardini, A.; Benedetti, M.G.; Berti, L.; Bettinelli, D.; Nativo, R.; Giannini, S. Rear-foot, mid-foot and fore-foot motion during the stance phase of gait. *Gait Posture* **2007**, *25*, 453–462. [CrossRef] [PubMed]
30. Shultz, R.; Jenkyn, T. Determining the maximum diameter for holes in the shoe without compromising shoe integrity when using a multi-segment foot model. *Med. Eng. Phys.* **2012**, *34*, 118–122. [CrossRef]
31. Sterzing, T.; Schweiger, V.; Ding, R.; Cheung, T.M.; Brauner, T. Influence of rearfoot and forefoot midsole hardness on biomechanical and perception variables during heel-toe running. *Footwear Sci.* **2013**, *5*, 71–79. [CrossRef]
32. Liu, Z.-L.; Lam, W.-K.; Zhang, X.; Vanwanseele, B.; Liu, H. Influence of heel design on lower extremity biomechanics and comfort perception in overground running. *J. Sport Sci.* **2021**, *39*, 232–238. [CrossRef] [PubMed]
33. Flores, N.; Rao, G.; Berton, E.; Delattre, N. The stiff plate location into the shoe influences the running biomechanics. *Sports Biomech.* **2021**, *20*, 815–830. [CrossRef] [PubMed]
34. Delattre, N.; Lafortune, M.A.; Moretto, P. Dynamic similarity during human running: About Froude and Strouhal dimensionless numbers. *J. Biomech.* **2009**, *42*, 312–318. [CrossRef]
35. Sloot, L.H.; Van der Krogt, M.M.; Harlaar, J. Energy exchange between subject and belt during treadmill walking. *J. Biomech.* **2014**, *47*, 1510–1513. [CrossRef] [PubMed]
36. Leardini, A.; Stebbins, J.; Hillstrom, H.; Caravaggi, P.; Deschamps, K.; Arndt, A. ISB recommendations for skin-marker-based multi-segment foot kinematics. *J. Biomech.* **2021**, *125*, 110581. [CrossRef]
37. Wu, G.; Siegler, S.; Allard, P.; Kirtley, C.; Leardini, A.; Rosenbaum, D.; Whittle, M.; D'Lima, D.D.; Cristofolini, L.; Witte, H.; et al. ISB recommendation on definitions of joint coordinate system of various joints for the reporting of human joint motion—Part I: Ankle, hip, and spine. *J. Biomech.* **2002**, *35*, 543–548. [CrossRef]
38. Cohen, J. *Statistical Power Analysis for the Behavioral Sciences*, 2nd ed.; Lawrence Erlbaum Associates: Mahwah, NJ, USA, 1988.
39. Altman, A.R.; Davis, I.S. A kinematic method for footstrike pattern detection in barefoot and shod runners. *Gait Posture* **2012**, *35*, 298–300. [CrossRef]

40. Cheung, R.T.H.; Ng, G.Y. A systematic review of running shoes and lower leg biomechanics: A possible link with patellofemoral pain syndrome? *Int. SportMed. J.* **2007**, *8*, 107–116. [CrossRef]
41. Kaufman, K.R.; Brodine, S.K.; Shaffer, R.; Johnson, C.W.; Cullison, T.R. The effect of foot structure and range of motion on musculoskeletal overuse injuries. *Am. J. Sports Med.* **1999**, *27*, 585. [CrossRef]
42. Willems, T.M.; Witvrouw, E.; De Cock, A.; De Clercq, D. Gait-related risk factors for exercise-related lower-leg pain during shod running. *Med. Sci. Sports Exerc.* **2007**, *39*, 330–339. [CrossRef]

Article

Shock Acceleration and Attenuation during Running with Minimalist and Maximalist Shoes: A Time- and Frequency-Domain Analysis of Tibial Acceleration

Liangliang Xiang [1,2], Yaodong Gu [1,2,*], Ming Rong [1,*], Zixiang Gao [1,3], Tao Yang [1], Alan Wang [2,4], Vickie Shim [2] and Justin Fernandez [2,5]

1   Faculty of Sports Science, Ningbo University, Ningbo 315211, China; lxia467@aucklanduni.ac.nz (L.X.); gaozixiang111@163.com (Z.G.); 1565409093@gmail.com (T.Y.)
2   Auckland Bioengineering Institute, The University of Auckland, Auckland 1010, New Zealand; alan.wang@auckland.ac.nz (A.W.); v.shim@auckland.ac.nz (V.S.); j.fernandez@auckland.ac.nz (J.F.)
3   Faculty of Engineering, University of Pannonia, H-8201 Veszprém, Hungary
4   Faculty of Medical and Health Sciences, The University of Auckland, Auckland 1010, New Zealand
5   Department of Engineering Science, Faculty of Engineering, The University of Auckland, Auckland 1010, New Zealand
*   Correspondence: guyaodong@nbu.edu.cn (Y.G.); rongming@nbu.edu.cn (M.R.); Tel.: +86-574-8760-9369 (Y.G.)

Abstract: Tibial shock attenuation is part of the mechanism that maintains human body stabilization during running. It is crucial to understand how shock characteristics transfer from the distal to proximal joint in the lower limb. This study aims to investigate the shock acceleration and attenuation among maximalist shoes (MAXs), minimalist shoes (MINs), and conventional running shoes (CONs) in time and frequency domains. Time-domain parameters included time to peak acceleration and peak resultant acceleration, and frequency-domain parameters contained lower (3–8 Hz) and higher (9–20 Hz) frequency power spectral density (PSD) and shock attenuation. Compared with CON and MAX conditions, MINs significantly increased the peak impact acceleration of the distal tibia ($p = 0.01$ and $p < 0.01$). Shock attenuation in the lower frequency depicted no difference but was greater in the MAXs in the higher frequency compared with the MIN condition ($p < 0.01$). MINs did not affect the tibial shock in both time and frequency domains at the proximal tibia. These findings may provide tibial shock information for choosing running shoes and preventing tibial stress injuries.

Keywords: running; impact loading; shock acceleration; shock attenuation; minimalist shoes; maximalist shoes

Citation: Xiang, L.; Gu, Y.; Rong, M.; Gao, Z.; Yang, T.; Wang, A.; Shim, V.; Fernandez, J. Shock Acceleration and Attenuation during Running with Minimalist and Maximalist Shoes: A Time- and Frequency-Domain Analysis of Tibial Acceleration. *Bioengineering* **2022**, *9*, 322. https://doi.org/10.3390/bioengineering9070322

Academic Editors: Rui Zhang, Wei-Hsun Tai and Ali Zarrabi

Received: 21 June 2022
Accepted: 14 July 2022
Published: 16 July 2022

Publisher's Note: MDPI stays neutral with regard to jurisdictional claims in published maps and institutional affiliations.

1. Introduction

Running is a prevalent worldwide form of exercise and with multiple benefits. However, runners suffer a high injury rate in their lower limbs [1,2]. Impact loading is a crucial measure to evaluate kinetic performance, and it is potentially associated with running-related injuries, such as patellofemoral pain and plantar fasciitis [3]. Ground reaction force (GRF) metrics, such as vertical peak GRF and vertical instantaneous load rate (VILR), are commonly employed to indicate impact loading during running. A greater VILR in runners has been associated with an increased risk of injury [3]. However, GRF metric measures usually need to be conducted in the laboratory setting with a force platform embedded in the running path or treadmill. Tibial acceleration has been shown to be strongly correlated with GRF metrics during running [4]. Moreover, collecting tibial acceleration from wearable inertial measurement unit (IMU) sensors is convenient outside the gait lab, cost-saving, and could increase the ecological validity.

Different running shoes are designed to decrease running-related lower limb injuries. Although modern shoes with cushioning are made to respect the natural foot shape and function evolutionarily, they may modify natural biomechanical characteristics during running [5]. Barefoot running has attracted lots of attention in the past decades. It can stimulate and strengthen inner foot muscles and maintain longitudinal arch function [6]. Inspired by barefoot running, minimalist shoes (MINs) aim to promote the natural movement of the foot and obtain barefoot-like biomechanical benefits during running, but without plantar surface injuries (i.e., blisters and bruises) [1]. MINs are characterized by a high flexibility, low weight, midsole stack height, and heel-to-toe drop without motion control and stability technologies/devices [5]. MIN running has been supported to promote foot function [7–10] and increase intrinsic foot muscle strength [7,11]. Nevertheless, rearfoot strike running in MINs may increase the impact loading rate, which has been suggested to increase the likelihood of injury in the shin and calf [12]. MIN running also presents a more pronounced instantaneous and average loading rate than conventional shoes (CONs) [13,14].

Maximalist shoes (MAXs) are distinguished by a high midsole stack height (typically greater than 30 mm) and excellent shock absorption properties [15,16]. They have been advertised in recent years with increased cushioning to protect runners from potential running-related injuries. Runners with MAXs may generate a smaller VILR than with MINs [12,13]. However, there are also some debates raised from recent studies. A previous study found that impact loading was increased after 5 km of running with MAXs [16]. MAXs may be unable to significantly decrease the impact loading metrics [17]. Nevertheless, limited studies have investigated tibial acceleration between MAXs and MINs [18–20].

Shock acceleration and attenuation characteristics in running may be affected by many factors. Peak tibial acceleration in the time and frequency domains was increased significantly for the habitual rearfoot runners than forefoot runners. Rearfoot runners also presented a significant shock attenuation effect in the lower and higher frequency ranges [21]. Lower limb impact attenuation is increased with the increase in step length during running [22]. Decreasing stride frequency resulted in a greater tibial impact acceleration and power spectral density (PSD) of the signal [23]. It was also found that peak tibial acceleration increased following a prolonged run, but shock attenuation did not change from a previous study [24]. During running, time- and frequency-domain features of the shock acceleration reflect the impact loading and attenuation functions of the footwear in the lower limbs. Sinclair et al. [13] demonstrated that runners exhibited higher tibial acceleration when wearing MINs than MAXs.

However, there is no compelling evidence supporting the differences in tibial shock acceleration in the time and frequency domains and how tibial shock attenuation changes between MAXs and MINs. Furthermore, recreational runners are frequently troubled by lower limb injuries, especially around the tibia, and the tibia absorbs a significant portion of the impact acceleration. Therefore, this study aimed to investigate the shock acceleration and attenuation in the tibia among MAXs, CONs, and MINs. We hypothesized that: (1) MINs increase and MAXs decrease peak shock compared to CONs in the time domain; (2) MIN running exhibits a significant shock attenuation effect due to it exhibiting a more natural barefoot running gait pattern from an evolutionary perspective and a greater power spectral magnitude at the distal tibia.

2. Materials and Methods

2.1. Participants

A minimum of twenty-one participants were required for this study (power: 0.8, effect size: 0.25, $\alpha = 0.05$ and $\beta = 0.2$). Therefore, this study recruited twenty-four male recreational runners (age: 28.3 ± 1.1 years, height: 1.76 ± 0.04 m, mass: 65.8 ± 2.2 kg, BMI: 21.3 ± 0.3 kg/m^2) from the university and local running clubs. Considering the differences in running mechanics between males and females [25], this study only focused on male participants. Inclusion criteria included recreational level runners, right leg-dominant runner, habitual rearfoot strike runners, and those who never had run in MINs or MAXs previously. The recreational runner was defined as running 2–4 times per week with a weekly running mileage no less than 20 km, and an age-graded score < 60th percentile calculated based on age, gender, and race performance in the past six months [26]. The rearfoot strike pattern was defined by the strike index (center of pressure within 0–33% of the foot length at the initial contact) using the Footscan® pressure plate (Rsscan International, Olen, Belgium) [27]. The test was conducted while participants were wearing the CONs. Exclusion criteria were as follows: BMI out of the range of 18–25, neurological or cardiovascular diseases, pes planus or pes cavus, and lower limb musculoskeletal injuries within the six months prior to participation in this study. All participants were free to exit the experiment at any test stage, and written informed consent was obtained from each participant before the test. The study protocol was conducted in compliance with the declaration of Helsinki and was approved by the University's Institutional Review Board (RAGH20201137).

2.2. Experiment Protocol

Participants were instructed to maintain their foot strike pattern during the test to avoid the effects of foot strike pattern change on the findings. Each participant was given ten minutes to run on the treadmill (Quasar, h/p cosmos®, GmbH, Germany) at a speed of 8 km/h for a warm-up and to become familiarized with the different shoes and experimental setting. All runners were required to wear a short running garment during the test. Tri-axial accelerometers (IMeasureU V1, Auckland, New Zealand; 40 × 28 × 15 cm, weight: 12 g, resolution: 16 bit) were attached on the proximal and distal anteromedial tibia of each participant's dominant leg using the strap, with the vertical axis aligning with the tibia (as shown in Figure 1a). All participants ran 6 min on the treadmill for each shoe condition with a speed of 10.8 ± 0.5 km/h, calculated according to the Froude velocity [28]. During the test, each subject ran 5 min to ensure that their gait stabilized and the last minute was record. The order for footwear selection was assigned randomly, and there was at least a ten-minute (10–30) break between each session to avoid fatigue effects [15,20]. The minimalist index was assessed among MINs, MAXs, and CONs, being 86%, 26%, and 36%, respectively (EURO sizes: 41–43); detailed information for each item is presented in Figure 1b. Each item was scored from 0 to 5. The minimalist index was evaluated based on an expert consensus from Esculier et al. [5] and was calculated by adding up all sub-scores, then multiplying by 0.04.

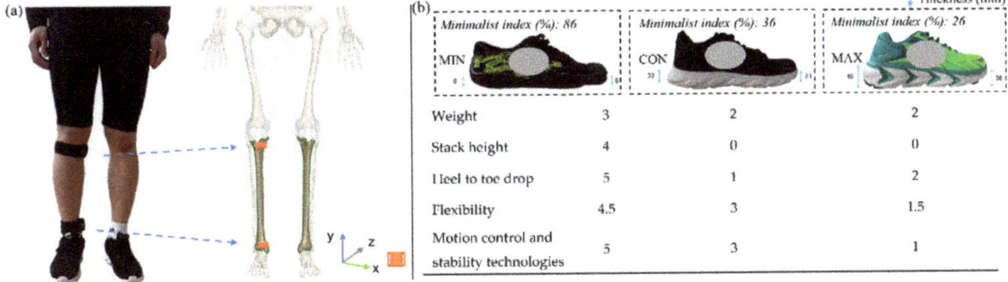

	MIN	CON	MAX
Weight	3	2	2
Stack height	4	0	0
Heel to toe drop	5	1	2
Flexibility	4.5	3	1.5
Motion control and stability technologies	5	3	1

Figure 1. The sensor placement (**a**) and the minimalist index and sub-scores for each item (**b**). Note: MINs: the minimalist shoes; CONs: the conventional shoes; MAXs: the maximalist shoes.

2.3. Data Collection and Processing

The tri-axis acceleration signal (Figure 2a) was sampled at 500 Hz and was filtered using a second-order, low-pass, zero-lag Butterworth filter with a cutoff frequency of 60 Hz to remove noise based on spectral analysis. Resultant accelerations were calculated as $\sqrt{x^2 + y^2 + z^2}$ (Figure 2b,c). We picked four steady gait cycles from each 10 s of the one-minute acceleration data according to the previously established and validated method, whereas the initial foot contact was the local minima within the 75 ms prior to the peak resultant distal tibial acceleration [29]. Therefore, each footwear condition resulted in 24 stance phases for time- and frequency-domain analysis. All parameters from the acceleration signal in this study were calculated using a custom Python program (v3.8, Python Software Foundation, Wilmington, NC, USA).

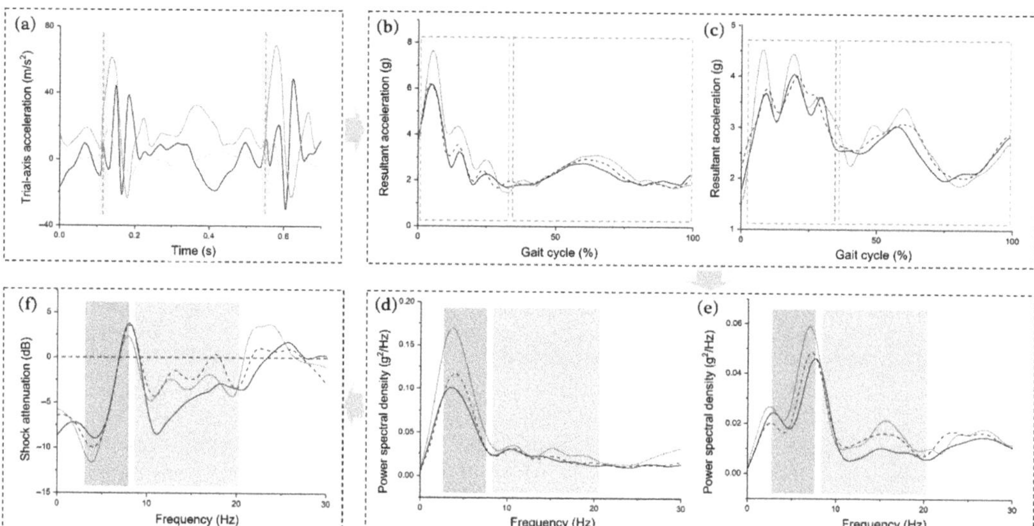

Figure 2. Graphical representation of the acceleration data process technique in the time and frequency domains. These include raw time-series tri-axis acceleration data (**a**); resultant acceleration from the distal (**b**) and proximal (**c**) tibia; power spectral density from the distal (**d**) and proximal (**e**) tibia; shock attenuation from the distal tibia to the proximal tibia (**f**). Note: black denotes maximalist shoes, gray denotes minimalist shoes, and the dashed line indicates conventional shoes.

The power spectrum was analyzed by transforming time-domain signals into the frequency domain using the discrete fast Fourier transform (FFT). A linear trend was removed by subtracting a least-squares best-fit line from the raw data signal [30]. Zero padding was performed at the end of each acceleration data until the total number of data points was 1024, as required by the FFT (a multiple of a power of two). The power of stance phase in the frequency domain (from 0 to the Nyquist frequency) was evaluated by calculating PSD using a rectangular window (Figure 2d,e). Furthermore, powers and frequencies were normalized to 1 Hz bins [31]. A transfer function [30] was employed to evaluate shock attenuation or gain between the distal and proximal tibia in decibels at each frequency interval using the following formula:

$$\text{Transfer Function} = 10 \log_{10}\left(PSD_{p\_tibia} / PSD_{d\_tibia}\right) \tag{1}$$

where PSD_{p_tibia} and PSD_{d_tibia} are the power spectral densities of the proximal and distal tibia. A positive value of the transfer function depicts a gain in signal strength at each frequency, and a negative value indicates the attenuation in signal power as the impact shock transfers from the proximal to the distal tibia (Figure 2f).

Therefore, time-domain parameters analyzed in this study included peak resultant acceleration and time from initial foot contact to peak resultant acceleration, and frequency-domain parameters contained lower (3–8 Hz) and higher (9–20 Hz) frequency PSD and shock attenuation.

2.4. Statistical Analysis

Prior to analysis, the Kolmogorov–Smirnov test was used to check the normality of data distribution, whereas homogeneity was assessed using Levene's test for homogeneity of variances. Greenhouse–Geisser corrected results are reported if data violated Mauchly's test for sphericity. A one-way repeated measures analysis of variance (ANOVA) was performed to determine the differences in time and frequency domains among MAXs, CONs, and MINs with a significance accepted at $p < 0.05$. The Bonferroni correction was used for the post hoc pairwise comparison with an adjusted significance level of $p < 0.017$. The effect size was evaluated to quantify the magnitude statistically using the partial eta-squared value (η_p^2) and classified as small ($0.01 < ES \leq 0.06$), medium ($0.06 < ES \leq 0.14$), and large ($ES > 0.14$) [32]. All statistical analyses were conducted using the SPSS v25 (IBM SPSS inc., Chicago, IL, USA) and GraphPad Prism 9.3.0 (San Diego, CA, USA) statistics software.

3. Results

No difference was presented for the time to peak acceleration (Table 1). The ANOVA analysis showed that peak resultant accelerations were different statistically among the three footwear conditions in both the distal and proximal tibia ($p < 0.01$, $\eta_p^2 = 0.4$ and $p = 0.01$, $\eta_p^2 = 0.2$). Compared with CON and MAX conditions, MINs significantly increased the peak acceleration magnitude of the distal tibia ($p = 0.01$ and $p < 0.01$) (Figure 3a). The peak acceleration in the MIN condition was also greater than the MAXs at the proximal tibia (5.7 ± 1.35 vs. 5.02 ± 0.9 g, $p < 0.01$) (Figure 3b).

Table 1. Tibial acceleration analysis in the time and frequency domains with different running shoes (data were presented in mean (SD)).

	MINs	CONs	MAXs	One-Way ANOVA F-Value	η_p^2	p-Value
Time domain						
Distal tibia						
Time to peak acceleration (s)	0.01 (0.00)	0.01 (0.00)	0.01 (0.01)	1.18	0.05	0.31
Peak resultant acceleration (g)	8.52 (1.75)	7.13 (1.37)	6.58 (0.91)	15.27	0.4	<0.01
Proximal tibia						
Time to peak acceleration (s)	0.03 (0.03)	0.05 (0.04)	0.06 (0.06)	2.01	0.08	0.15
Peak resultant acceleration (g)	5.7 (1.35)	5.32 (1.10)	5.02 (0.90)	5.73	0.2	0.01
Frequency domain						
Distal tibia						
PSD in 3–8 Hz (g^2/Hz)	0.68 (0.14)	0.48 (0.18)	0.42 (0.14)	18.99	0.45	<0.01
PSD in 9–20 Hz (g^2/Hz)	0.32 (0.14)	0.26 (0.14)	0.25 (0.10)	2.89	0.11	0.07
Proximal tibia						
PSD in 3–8 Hz (g^2/Hz)	0.23 (0.12)	0.19 (0.10)	0.18 (0.10)	4.30	0.16	0.02
PSD in 9–20 Hz (g^2/Hz)	0.17 (0.10)	0.16 (0.08)	0.12 (0.07)	6.83	0.23	<0.01
Shock attenuation						
3–8 Hz magnitude (dB)	−32.36 (21.28)	−28.12 (23.12)	−24.61 (23.76)	1.98	0.08	0.15
9–20 Hz magnitude (dB)	−38.27 (45.03)	−23.53 (42.64)	−54.72 (37.49)	5.40	0.19	0.01

Note: PSD: power spectral density; MINs: the minimalist shoes; CONs: the conventional shoes; MAXs: the maximalist shoes.

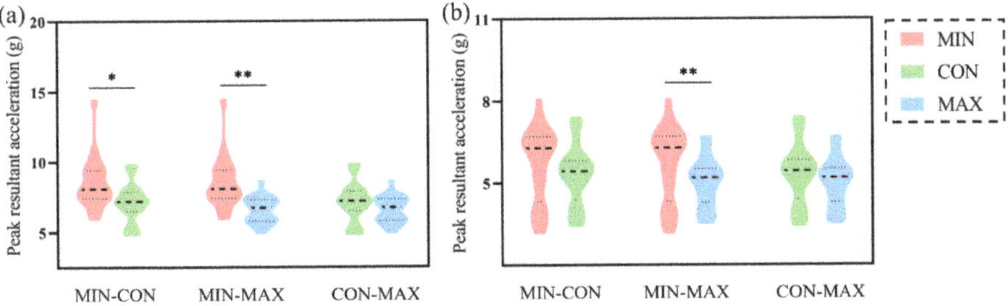

Figure 3. Violin plots of the Bonferroni comparisons for peak resultant acceleration between conditions in the distal (**a**) and proximal (**b**) tibia. Note: MINs: the minimalist shoes; CONs: the conventional shoes; MAXs: the maximalist shoes. * $p < 0.05$ and ** $p < 0.01$. The black dashed line represents the median, and the gray dashed lines above and below represent the third and first quartiles.

In the distal tibia, PSD in the lower frequency (3–8 Hz) exhibited statistical differences among the three conditions with $\eta_p^2 = 0.45$ and $p < 0.01$ (Table 1), and was greater in the MIN condition than the CON ($p < 0.01$) and MAX ($p < 0.01$) conditions (Figure 4a). In the proximal tibia, it was demonstrated that MAXs decreased the PSD in both the lower ($p = 0.03$) and higher ($p < 0.01$) frequency range compared to MINs (Figure 4b,c). PSD in the higher frequency was also less in the statistics than CONs (0.12 ± 0.07 vs. 0.16 ± 0.08 g^2/Hz, $p = 0.02$). Shock attenuation in the lower frequency depicted no difference but was greater in the MAXs in the higher frequency (9–20 Hz) compared with the MIN condition (-54.72 ± 37.49 vs. -38.27 ± 45.03 dB and $p < 0.01$).

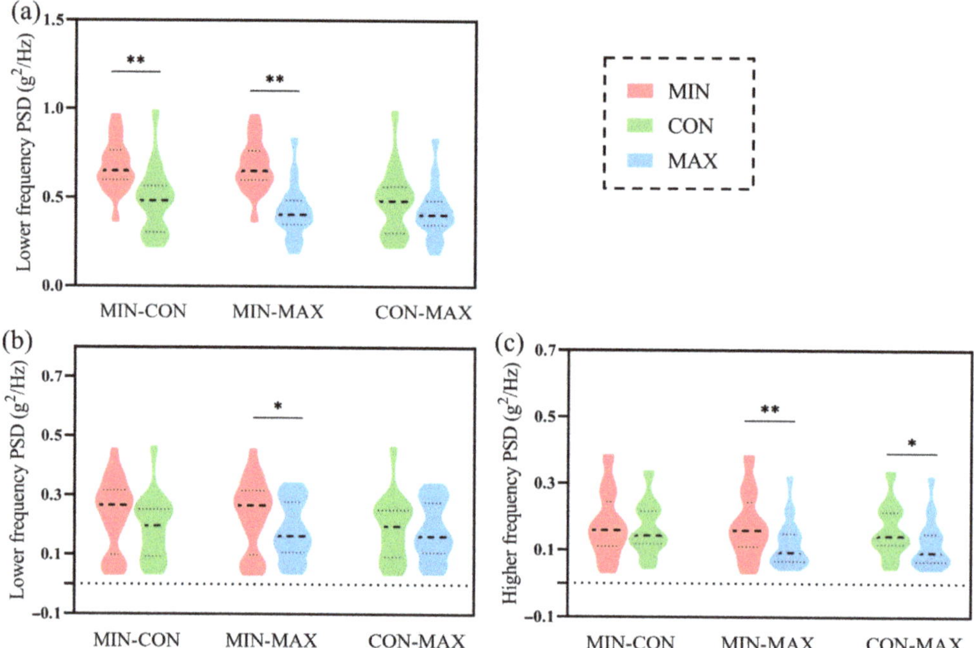

Figure 4. Violin plots of the Bonferroni comparisons for PSD between conditions in the distal (**a**) and proximal (**b**,**c**) tibia. Note: PSD: power spectral density; MINs: the minimalist shoes; CONs: the conventional shoes; MAXs: the maximalist shoes. * $p < 0.05$ and ** $p < 0.01$. The black dashed line represents the median, and the gray dashed lines above and below represent the third and first quartiles.

4. Discussion

Time- and frequency-domain characteristics were investigated in this study among the MINs, MAXs, and CONs. We found that tibial shock acceleration differs among the different footwear conditions in the time domain and altered PSD and shock attenuation in the frequency dimension. Specifically, MINs increased peak acceleration in the distal tibia, but peak acceleration was not significantly different in the proximal tibia. PSD in the MINs was increased at the distal tibia in the lower frequency. MAXs decreased PSD on the proximal tibia (9–20 Hz). Furthermore, PSD in the MINs was significantly greater than in the MAXs at both the lower and higher frequency ranges in the proximal tibia and the lower frequency range in the distal tibia.

Shock acceleration characteristics in time and frequency dimensions are essential to understand impact loading. Footwear [13] and prolonged running [24] are associated with peak acceleration alterations. The frequency content of the impact loading and how shock acceleration is attenuated are thought to be of greater importance for understanding injury mechanisms and preventing potential injuries than variables in the time domain [21]. Foot strike pattern [21], step length [22], and stride frequency [23] have been previously identified as potential factors that contribute to differences in PSD during running. This study illustrated the differences in the frequency characteristics between MINs and MAXs. Furthermore, MIN running lacks the cushioning function and MAX running decreases impact loading metrics in the time domain, which has been controversial [15,17]. We found that MAXs did not decrease impact acceleration significantly in either the time or frequency domains, except for PSD at the higher frequency in the proximal tibia, which is in contrast with our hypothesis.

Sinclair [19] evaluated shock attenuation and illustrated decreased impact attenuation in the MINs compared to MAXs and CONs. However, it was quantified in the whole lower limb from a time dimension [33]. Shin and knee injuries are predominant in running-related injuries, for instance, patellofemoral pain [34] and bone stress injuries [35]. It is essential to explore the transmission of impact loading from the ankle and distal tibia to the proximal tibia and knee. We could identify that the cushioning function affects the time to peak acceleration, as it was decreased at the proximal tibia in the MINs but increased in the MAXs, although with no significant effect. Therefore, the impact loading rate may be increased in the MINs and decreased in the MAXs as the peak acceleration at the proximal tibia was higher in the MINs, followed by the CONs, which is consistent with the previous finding [36].

The pathomechanical evidence from a previous literature review [35] supports that this enhanced impulse may increase the risk of bone stress injuries in distance runners. The difference in footwear in shock acceleration is present in both the distal and proximal tibia and may be a contributing factor to bone stress fractures. It also appeared that peak acceleration in the MINs was greater than in the CONs at the initial tibia reaction, but it was not significant when the impact acceleration reached the knee, which is consistent with the frequency characteristic in the lower range frequency components. Hence, the cushioning function from footwear in the time domain is more pronounced from the ankle to the distal tibia and dissipates at the proximal tibia. These findings elucidated the shock absorption mechanism at the tibia among different footwear conditions. A previous study from Busa et al. [23] suggested defining the lower frequency signal as the active phase of stance and the higher frequency content as the impact phase; however, impact and active phases were not set to these specific ranges because the Fourier transform loses all time-domain information [30].

Additionally, the PSD difference from the distal to the proximal tibia and shock attenuation characteristics in this study supported that quantifying the shock attenuation at the tibia is critical for understanding acceleration and impact absorption differences among footwear conditions in the time and frequency domains [19,37] because impact loading maintains a similar magnitude from the knee [38].

The greater the shock attenuation, the more impact loading is dissipated by the tibia during the stance phase [21]. In support of our hypothesis, MINs depicted a greater shock attenuation effect in the higher frequency domain than CONs. However, only MAXs showed a significant increase in the shock attenuation. This is inconsistent with our expectation that MINs would exhibit a more considerable shock attenuation than MAXs, compared to the CON condition. It means the cushioning function in the MAXs presented the importance of absorbing shock regarding the frequency content. Therefore, the shock attenuation function in the MAXs exist as the footwear manufacturer claimed, rather than just as a commercial advert. Investigating the time and frequency contents for the MAXs goes beyond understanding the impact loading during running in just the time domain [15,20] and sheds light on the understanding of shock acceleration in the frequency content. These findings have potential clinical implications, specifically for preventing tibial stress injuries in recreational male runners using wearable sensors.

One limitation in this study should be considered, namely, only male habitual rearfoot strike runners were recruited in this study. Therefore, these findings are not suggested to apply for the habitual mid- and forefoot strike runner. Future studies may be interested in investigating shock attenuation characteristics on forefoot strikers in the time and frequency contents between MIN and MAX conditions. Differences between male and female runners may also reveal differences in the injury mechanics in the lower extremities, which is worth exploring in the future.

5. Conclusions

This study is a timely addition to the literature regarding the time-domain tibial shock and frequency-domain shock attenuation between the distal and proximal tibia in male recreational runners between MIN, MAX, and CON conditions. We found that peak acceleration and PSD at the distal tibia were significantly greater in the MINs than CONs during running, but the difference disappeared when the impact loading transferred to the proximal tibia. However, MINs demonstrated no significant shock attenuation effect. These findings may provide tibial shock information for choosing running shoes and preventing tibial stress injuries. It is suggested that novice runners and recreational runners with a history of tibial stress fractures do not use the MINs as their preference in daily running.

Author Contributions: Conceptualization, L.X. and Y.G.; methodology, L.X., Z.G., M.R. and T.Y.; software, L.X.; validation, A.W., V.S. and J.F.; formal analysis, L.X., M.R. and T.Y.; investigation, L.X. and V.S.; resources, L.X. and M.R.; data curation, L.X.; writing—original draft preparation, L.X.; writing—review and editing, A.W. and J.F.; visualization, Z.G.; supervision, Y.G., A.W., V.S. and J.F.; project administration, Y.G.; funding acquisition, Y.G. and M.R.; All authors have read and agreed to the published version of the manuscript.

Funding: This research was supported by the Key R&D Program of Zhejiang Province, China, grant number 2021C03130; Public Welfare Science and Technology Project of Ningbo, China, grant number 2021S133; Zhejiang Province Science Fund for Distinguished Young Scholars, grant number R22A021199; and K. C. Wong Magna Fund of Ningbo University. Liangliang Xiang is being sponsored by the China Scholarship Council (CSC).

Institutional Review Board Statement: The study was conducted in accordance with the Declaration of Helsinki, and approved by the Institutional Review Board of Faculty of Sports Science, Ningbo University (protocol code: RAGH20201137 and date of approval: 21 January 2021).

Informed Consent Statement: Informed consent was obtained from all subjects involved in the study.

Data Availability Statement: Data is available on request due to the restriction of ethics.

Conflicts of Interest: The authors declare no conflict of interest.

References

1. Altman, A.R.; Davis, I.S. Prospective comparison of running injuries between shod and barefoot runners. *Br. J. Sports Med.* **2015**, *50*, 476–480. [CrossRef] [PubMed]
2. Murr, S.; Pierce, B. How Aging Impacts Runners' Goals of Lifelong Running. *Phys. Act. Health* **2019**, *3*, 71–81. [CrossRef]
3. Johnson, C.D.; Tenforde, A.S.; Outerleys, J.; Reilly, J.; Davis, I.S. Impact-Related Ground Reaction Forces Are More Strongly Associated With Some Running Injuries Than Others. *Am. J. Sports Med.* **2020**, *48*, 3072–3080. [CrossRef] [PubMed]
4. Tenforde, A.S.; Hayano, T.; Jamison, S.T.; Outerleys, J.; Davis, I.S. Tibial Acceleration Measured from Wearable Sensors Is Associated with Loading Rates in Injured Runners. *PM&R* **2020**, *12*, 679–684. [CrossRef]
5. Esculier, J.-F.; Dubois, B.; Dionne, C.E.; Leblond, J.; Roy, J.-S. A consensus definition and rating scale for minimalist shoes. *J. Foot Ankle Res.* **2015**, *8*, 42. [CrossRef]
6. Lieberman, D.E. What We Can Learn About Running from Barefoot Running: An evolutionary medical perspective. *Exerc. Sport Sci. Rev.* **2012**, *40*, 63–72. [CrossRef]
7. Miller, E.E.; Whitcome, K.K.; Lieberman, D.E.; Norton, H.L.; Dyer, R.E. The effect of minimal shoes on arch structure and intrinsic foot muscle strength. *J. Sport Health Sci.* **2014**, *3*, 74–85. [CrossRef]
8. Xiang, L.; Mei, Q.; Wang, A.; Shim, V.; Fernandez, J.; Gu, Y. Evaluating function in the hallux valgus foot following a 12-week minimalist footwear intervention: A pilot computational analysis. *J. Biomech.* **2022**, *132*, 110941. [CrossRef]
9. Xiang, L.; Mei, Q.; Wang, A.; Fernandez, J.; Gu, Y. Gait biomechanics evaluation of the treatment effects for hallux valgus patients: A systematic review and meta-analysis. *Gait Posture* **2022**, *94*, 67–78. [CrossRef]
10. Xiang, L.; Mei, Q.; Fernandez, J.; Gu, Y. Minimalist shoes running intervention can alter the plantar loading distribution and deformation of hallux valgus: A pilot study. *Gait Posture* **2018**, *65*, 65–71. [CrossRef]

11. Fuller, J.T.; Thewlis, D.; Tsiros, M.D.; Brown, N.A.T.; Hamill, J.; Buckley, J.D. Longer-term effects of minimalist shoes on running performance, strength and bone density: A 20-week follow-up study. *Eur. J. Sport Sci.* **2018**, *19*, 402–412. [CrossRef] [PubMed]

12. Hannigan, J.; Pollard, C.D. Differences in running biomechanics between a maximal, traditional, and minimal running shoe. *J. Sci. Med. Sport* **2020**, *23*, 15–19. [CrossRef] [PubMed]

13. Sinclair, J.; Richards, J.; Selfe, J.; Fau-Goodwin, J.; Shore, H. The Influence of Minimalist and Maximalist Footwear on Patellofemoral Kinetics During Running. *J. Appl. Biomech.* **2016**, *32*, 359–364. [CrossRef] [PubMed]

14. Willy, R.; Davis, I.S. Kinematic and Kinetic Comparison of Running in Standard and Minimalist Shoes. *Med. Sci. Sports Exerc.* **2014**, *46*, 318–323. [CrossRef] [PubMed]

15. Chan, Z.Y.S.; Au, I.P.H.; Lau, F.O.Y.; Ching, E.C.K.; Zhang, J.H.; Cheung, R.T.H. Does maximalist footwear lower impact loading during level ground and downhill running? *Eur. J. Sport Sci.* **2018**, *18*, 1083–1089. [CrossRef] [PubMed]

16. Pollard, C.D.; Ter Har, J.A.; Hannigan, J.J.; Norcross, M.F. Influence of Maximal Running Shoes on Biomechanics Before and After a 5K Run. *Orthop. J. Sports Med.* **2018**, *6*, 1–5. [CrossRef]

17. Hannigan, J.; Pollard, C.D. A 6-Week Transition to Maximal Running Shoes Does Not Change Running Biomechanics. *Am. J. Sports Med.* **2019**, *47*, 968–973. [CrossRef]

18. Agresta, C.; Kessler, S.; Southern, E.; Goulet, G.C.; Zernicke, R.; Zendler, J. Immediate and short-term adaptations to maximalist and minimalist running shoes. *Footwear Sci.* **2018**, *10*, 95–107. [CrossRef]

19. Sinclair, J. The influence of minimalist, maximalist and conventional footwear on impact shock attenuation during running. *Mov. Sport Sci. Sci. Mot.* **2016**, *95*, 59–64. [CrossRef]

20. Mo, S.; Chan, Z.Y.S.; Lai, K.K.Y.; Chan, P.P.-K.; Wei, R.X.-Y.; Yung, P.S.-H.; Shum, G.; Cheung, R.T.-H. Effect of minimalist and maximalist shoes on impact loading and footstrike pattern in habitual rearfoot strike trail runners: An in-field study. *Eur. J. Sport Sci.* **2020**, *21*, 183–191. [CrossRef]

21. Gruber, A.H.; Boyer, K.A.; Derrick, T.; Hamill, J. Impact shock frequency components and attenuation in rearfoot and forefoot running. *J. Sport Health Sci.* **2014**, *3*, 113–121. [CrossRef]

22. Baggaley, M.; Vernillo, G.; Martinez, A.; Horvais, N.; Giandolini, M.; Millet, G.Y.; Edwards, W.B. Step length and grade effects on energy absorption and impact attenuation in running. *Eur. J. Sport Sci.* **2019**, *20*, 756–766. [CrossRef] [PubMed]

23. Busa, M.A.; Lim, J.; Van Emmerik, R.E.A.; Hamill, J. Head and Tibial Acceleration as a Function of Stride Frequency and Visual Feedback during Running. *PLoS ONE* **2016**, *11*, e0157297. [CrossRef] [PubMed]

24. Reenalda, J.; Maartens, E.; Buurke, J.H.; Gruber, A.H. Kinematics and shock attenuation during a prolonged run on the athletic track as measured with inertial magnetic measurement units. *Gait Posture* **2018**, *68*, 155–160. [CrossRef]

25. Ferber, R.; Davis, I.M.; Williams, D.S. Gender differences in lower extremity mechanics during running. *Clin. Biomech.* **2003**, *18*, 350–357. [CrossRef]

26. Liu, Q.; Mo, S.; Cheung, V.C.; Cheung, B.M.; Wang, S.; Chan, P.P.; Malhotra, A.; Cheung, R.T.; Chan, R.H. Classification of runners' performance levels with concurrent prediction of biomechanical parameters using data from inertial measurement units. *J. Biomech.* **2020**, *112*, 110072. [CrossRef]

27. Altman, A.R.; Davis, I.S. A kinematic method for footstrike pattern detection in barefoot and shod runners. *Gait Posture* **2012**, *35*, 298–300. [CrossRef]

28. England, S.A.; Granata, K.P. The influence of gait speed on local dynamic stability of walking. *Gait Posture* **2007**, *25*, 172–178. [CrossRef]

29. Aubol, K.G.; Milner, C. Foot contact identification using a single triaxial accelerometer during running. *J. Biomech.* **2020**, *105*, 109768. [CrossRef]

30. Shorten, M.R.; Winslow, D.S. Spectral Analysis of Impact Shock during Running. *Int. J. Sport Biomech.* **1992**, *8*, 288–304. [CrossRef]

31. Hamill, J.; Derrick, T.; Holt, K. Shock attenuation and stride frequency during running. *Hum. Mov. Sci.* **1995**, *14*, 45–60. [CrossRef]

32. Cohen, J. *Statistical Power Analysis for the Behavioral Sciences*, 2nd ed.; Routledge: New York, NY, USA, 2013.

33. García-Pérez, J.A.; Pérez-Soriano, P.; Belloch, S.L.; Lucas-Cuevas, Á.G.; Sánchez-Zuriaga, D. Effects of treadmill running and fatigue on impact acceleration in distance running. *Sports Biomech.* **2014**, *13*, 259–266. [CrossRef] [PubMed]

34. Willwacher, S.; Kurz, M.; Robbin, J.; Thelen, M.; Hamill, J.; Kelly, L.; Mai, P. Running-Related Biomechanical Risk Factors for Overuse Injuries in Distance Runners: A Systematic Review Considering Injury Specificity and the Potentials for Future Research. *Sports Med.* **2022**; Advance online publication. [CrossRef]

35. Hoenig, T.; Ackerman, K.E.; Beck, B.R.; Bouxsein, M.L.; Burr, D.B.; Hollander, K.; Popp, K.L.; Rolvien, T.; Tenforde, A.S.; Warden, S.J. Bone stress injuries. *Nat. Rev. Dis. Prim.* **2022**, *8*, 26. [CrossRef] [PubMed]

36. Dempster, J.; Dutheil, F.; Ugbolue, U.C. The Prevalence of Lower Extremity Injuries in Running and Associated Risk Factors: A Systematic Review. *Phys. Act. Health* **2021**, *5*, 133–145. [CrossRef]

37. Schütte, K.H.; Seerden, S.; Venter, R.; Vanwanseele, B. Influence of outdoor running fatigue and medial tibial stress syndrome on accelerometer-based loading and stability. *Gait Posture* **2018**, *59*, 222–228. [CrossRef]

38. McErlain-Naylor, S.; King, M.; Allen, S. Surface acceleration transmission during drop landings in humans. *J. Biomech.* **2021**, *118*, 110269. [CrossRef]

Article

Effect of Heel Lift Insoles on Lower Extremity Muscle Activation and Joint Work during Barbell Squats

Zhenghui Lu [1], Xin Li [1], Rongrong Xuan [2,*], Yang Song [3], István Bíró [3], Minjun Liang [1,*] and Yaodong Gu [1]

[1] Faculty of Sports Science, Ningbo University, Ningbo 315211, China; luzhenghui_nbu@foxmail.com (Z.L.); 2011042028@nbu.edu.cn (X.L.); guyaodong@nbu.edu.cn (Y.G.)
[2] The Affiliated Hospital of Medical School of Ningbo University, Ningbo 315020, China
[3] Faculty of Engineering, University of Szeged, 6720 Szeged, Hungary; nbusongyang@hotmail.com (Y.S.); biro-i@mk.u-szeged.hu (I.B.)
* Correspondence: fyxuanrongrong@nbu.edu.cn (R.X.); liangminjun@nbu.edu.cn (M.L.)

Abstract: The effect of heel elevation on the barbell squat remains controversial, and further exploration of muscle activity might help find additional evidence. Therefore, 20 healthy adult participants (10 males and 10 females) were recruited for this study to analyze the effects of heel height on lower extremity kinematics, kinetics, and muscle activity using the OpenSim individualized musculoskeletal model. One-way repeated measures ANOVA was used for statistical analysis. The results showed that when the heel was raised, the participant's ankle dorsiflexion angle significantly decreased, and the percentage of ankle work was increased ($p < 0.05$). In addition, there was a significant increase in activation of the vastus lateralis, biceps femoris, and gastrocnemius muscles and a decrease in muscle activation of the anterior tibialis muscle ($p < 0.05$). An increase in knee moments and work done and a reduction in hip work were observed in male subjects ($p < 0.05$). In conclusion, heel raises affect lower extremity kinematics and kinetics during the barbell squat and alter the distribution of muscle activation and biomechanical loading of the joints in the lower extremity of participants to some extent, and there were gender differences in the results.

Keywords: barbell squat; OpenSim; lower limb biomechanics; muscle; joint work

Citation: Lu, Z.; Li, X.; Xuan, R.; Song, Y.; Bíró, I.; Liang, M.; Gu, Y. Effect of Heel Lift Insoles on Lower Extremity Muscle Activation and Joint Work during Barbell Squats. *Bioengineering* **2022**, *9*, 301. https://doi.org/10.3390/bioengineering9070301

Academic Editors: Rui Zhang, Wei-Hsun Tai and Ali Zarrabi

Received: 31 May 2022
Accepted: 5 July 2022
Published: 8 July 2022

Publisher's Note: MDPI stays neutral with regard to jurisdictional claims in published maps and institutional affiliations.

1. Introduction

The barbell squat is one of the most effective exercises for building lower extremity strength [1–15] and is widely used for strength training and rehabilitation. During barbell squat training or competition, athletes often wear a type of heel lift shoe or use other means to elevate the heel, which is thought to improve the range of motion (ROM) of the lower extremity joints and improve stability of movement during the deep squat [16–21]. However, there is still controversy about whether raising the heel will affect the barbell squat movement.

Previous studies have shown that elevating heel height while performing barbell squats reduces the ankle dorsiflexion angle and anterior trunk tilt [16,18,22–24], which might help reduce lumbar spine shear forces [25]. However, some studies have reported that heel elevation does not affect hip and knee kinematic parameters [6,23,24]. In a study by Lee et al. [6], there were no significant effects on trunk and knee kinematics, whether using an inclined platform or weightlifting shoes. The participants recruited in the study by Lee et al. were all recreational weight lifters [6], and their movement patterns when performing a high load (80% 1RM) barbell squat could have large individual differences. Moreover, in the study by Lee et al. [6], participants were asked to squat until their hips reached the level of their knees rather than reaching maximum depth, which may have resulted in some participants not having reached the full ROM that they could accomplish. Furthermore, it is worth noting that although the variability in results was not significant, Lee et al. still observed greater knee flexion angles when raising the heel. In a recently published review [11], Pangan et al. noted that most current studies have focused on the effects of heel elevation on the kinematics of the deep squat movement, and fewer studies have examined the effects on kinetics, which may provide some valuable evidence for these effects.

Although coaches and athletes have been hoping to improve the movement pattern of the barbell squat by elevating the heel, the human neuromuscular system has the specific ability to adapt and coordinate to changes in the external environment, and therefore, small heel changes might hardly have a significant effect on the movement. Perhaps, it is reasonable to speculate that in some studies [6], lifters with some experience already have a relatively regular movement pattern. When the heel is elevated, the trainer would actively adjust the joint activity to fit the change, thus, the kinematic effects of elevated heel height on the deep squat may be difficult to observe. This was also demonstrated in a study by Sayers et al. [25]. Therefore, it might be possible to provide additional insight by studying kinetic parameters and muscle activity.

During the barbell squat, it is often necessary to engage the joint with the optimal moment and angular velocity to perform the barbell squat better [26–28]. The output power is closely related to the coordination of forces [26], rather than a solely single muscle [29]. When the heel is raised, it might cause the angle and ROM of the lower extremity joints to change, which may modify the joint's mechanical forces and affect the joint's work. In addition, during the barbell squat, especially during descent, some lower limb muscles are actively engaged in eccentric contraction to maintain the stability of the movement, involving energy dissipation, which is essential for the control of the action. Currently, many studies have used energetic principles to investigate the work done by joints in various states of motion [30–34]. These studies often use line and bar graphs to produce reports of the work done by the joints, which may be difficult to interpret with a single overlapping line and error line [35]. Recently, Ebrahimi et al. proposed composite lower limb work (CLEW) [35], a visualization method for interpreting lower limb work that has been used for landing, jumping, and gait tasks [35–37]. However, little attention has been paid to the effect of heel height on lower limb work during the barbell squat. In conclusion, the effect of elevated heel height on the kinematics and kinetics of the barbell deep squat remains controversial due to the differences in study designs. Pangan et al. suggested [11] that relevant studies could consider the gender and analyze muscle activity, since joint mobility and anatomical differences between men and women may affect kinematics and kinetics [23,24], and exploring muscle activity may help to explain the test results.

As a multi-joint exercise, the barbell squat primarily requires activating the hip, knee, and ankle joints [11]. Several studies have used surface electromyography (EMG) to test the effects of different foot conditions on the barbell squat [23,25,38]. Anbarian et al. found that raising the heel reduced fatigue in the lower back muscles [39]. Johnston et al. found that raising the heel significantly increased muscle activation in the gastrocnemius [40]. However, it is difficult to obtain information about deeper muscles using surface EMG, and the fat between the skin and the muscle may affect the measurements. OpenSim is an open-source biomechanical modeling, simulation, and analysis software. It enables the calculation of muscle activation by minimizing the sum of squares of muscle activation, and by using simulations, it can be used for calculating the degree of activation of deep muscles.

Therefore, in this study, we used an OpenSim musculoskeletal model specifically for the deep squat to reveal the effects of insoles with different heel heights on lower extremity muscle activation and work done during the barbell deep squat [8]. Based on previous studies, we hypothesized that an elevated heel would reduce mobility, extension moments, and joint work reduction in the hip and ankle joints instead of the knee.

2. Materials and Methods

2.1. Participants

The sample size for the within-factor repeated measures ANOVA (1 group × 3 measurements) was calculated using G*Power 3.1 (Franz Faul, Germany), as suggested by

Faul et al. [41], taking into consideration Cohen's medium effect size (d = 0.5), α error probability = 0.05, and power (1 − β) = 0.8. Based on these parameters, it was estimated that a minimum of 9 participants would be required for this study.

Twenty participants (female/male, 10/10) were eventually recruited for the experiment. All the participants were healthy and had more than 12 months of strength training experience, and the 1RM load of the barbell squat was between 1 and 1.2 times body weight for the female participants and between 1.75 and 2 times body weight for the male participants. They had no history of back and lower limb pain or injury in the most recent year and did not have any exercise in the 72 h before the experiment. All participants were aware of the test's purpose, method, and steps and filled in the informed consent form. The study was approved by the Ethics Committee of Ningbo University. Table 1 summarizes the demographic information related to the age, height, and weight of the participants.

Table 1. Basic information of participants.

Index	Gender	
	Male	Female
n	10	10
Age (years)	24.00 ± 1.48	23.50 ± 1.12
Height (m)	1.75 ± 0.03	1.66 ± 0.02
Weight (kg)	79.00 ± 5.34	59.25 ± 3.96
Training age (years)	2.00 ± 0.71	2.75 ± 1.09

Participants' height and weight tests were performed uniformly before the experiment.

2.2. Experimental Design

Reflective point trajectories were captured using a Vicon 3D motion capture system (v. 1.8.5, Vicon Motion System, Oxford, UK) with eight cameras (MX-T-Series) at 200 Hz and an AMTI 3D force meter (AMTI, Watertown, MA, USA) at 1000 Hz to capture the marker point trajectory and ground reaction force simultaneously (Figure 1c,d). At least two cameras could capture the reflective markers during the entire barbell squat movement, avoiding gaps during data collection or marker points in a particular direction obscuring the barbell, therefore, ensuring accuracy.

The surface hair, body surface grease, sweat, and other foreign matter that may affect EMG signal collection were removed with a razor before attaching the sensor. The sensor was attached to the muscle belly of the target muscle, and the electrodes were placed parallel to the direction of the muscle fibers, with the direction of the muscle fibers referenced to the bony markers on the participant's body surface. Using the rectus femoris muscle as an example, the electrodes need to be placed at 50% on the line from the anterior spina iliaca superior to the superior part of the patella. The EMG signals of the rectus femoris, biceps femoris longus, tibialis anterior muscle, and gastrocnemius muscle were collected synchronously with 1000 Hz frequency using a wireless Delsys EMG test system (Delsys, Boston, MA, USA) for model verification. The maximum voluntary contraction (MVC) was measured using a CON-TREX dynamometer (CON-TREX MJ System, CMV, Dübendorf, Switzerland). The original EMG signal was first filtered by bandpass fourth-order Butterworth filter in the frequency range of 100–500 Hz in Delsys EMG works analysis software. The amplitude analysis is carried out using root mean square (RMS) calculation. The MVC and standardized activity value of each action was output. The EMG activity was calculated from 0 (0, completely inactive) to 1 (100%, fully activated) through the test root mean square amplitude/MVC root mean square amplitude.

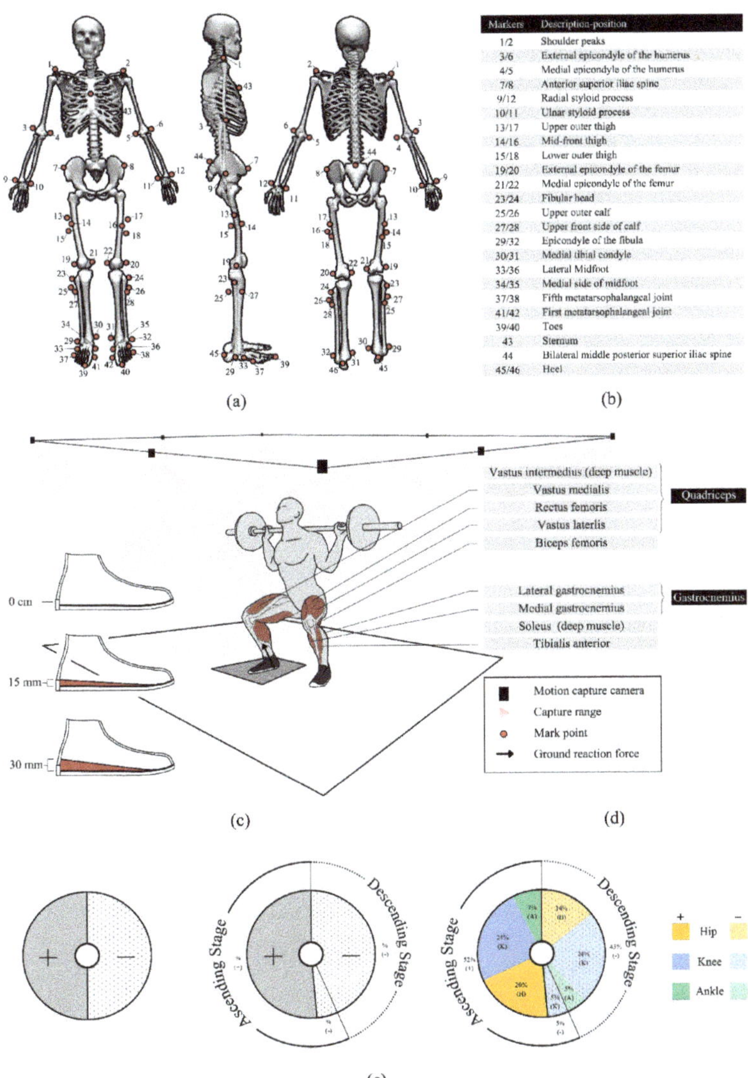

Figure 1. (**a**) Marking point paste location; (**b**) description of marker point location; (**c**) lab environment, barbell squatting movements, major muscle groups, and heel life insoles; (**d**) major muscle descriptions and illustrations; (**e**) Creation of a CLEW method pie chart for barbell squats; (**e-left**) first, the pie chart is divided according to the positive and negative work done. Starting from the centerline, the positive work is on the left side of the circle (dark), while the negative work is on the right side of the process (light dot); (**e-middle**) secondly, the pie chart is divided according to the descending and ascending stages of the squatting movement. The dotted part of the outer circle indicates the descending phase of the squatting movement, and the solid line represents the ascending stage. The percentage of the relative positive or negative work in the absolute stage work is expressed in the outer ring; (**e-right**) finally, each relevant constituent work is designated as part of the pie chart, squatting down in the order of the hips, knees, and ankles, starting with the central reference line, and the colors and patterns inside the pie chart are used to distinguish joints and to separate positive work from negative work.

All participants performed a deep squat weight limit test under the guidance of a relevant practitioner. Participants performed a 5 min squat exercise to warm up before the experiment. In this study, only the dominant foot of the participants was analyzed. The dominant foot was determined using the customary soccer kick. All participants involved in the experiment had the right side as the dominant foot. All participants wore uniformly provided tights during the test, and the marking points were taped by the same skilled tester (Figure 1a,b). The calibration procedure of the VICON 3D motion capture system was performed according to the recommendations of the camera manufacturer. Each participant performed barbell squats in a randomized order wearing shoes with 0 cm, 1.5 cm, and 3 cm heel inserts (Figure 1c). The participants could not be blinded because the participants perceived the intervention conditions. Each participant used a high bar position squat with a weight selection of 70% 1RM and the right foot in contact with the force platform during the test. The 70% 1RM load is considered to be a typical choice of training load [23] and is also considered to have low movement variability and high retest reliability [42–45] to maintain the level of repeatability required to obtain a representative dataset. The test began with the knees in an upright position, squatting down to the deepest point while ensuring a neutral pelvic position and keeping the heels off the ground. Then, participants extended the hips, knees, and ankles, and returned to the upright position and the end of the movement. There was at least 2 min between each deep squat, and then participants rated their perceived exertion using a scale to ensure participants were well rested. All variables (min, max, and ROM) were obtained at each squat, and then averaged for each group. Finally, three sets of data for each heel height condition were used for statistical purposes.

2.3. Musculoskeletal Model and Model Validation

A customized deep squat model was used in this study [8,46]. This model has been used several times in previous studies [8,47]. The model was validated using the standardized muscle electrical signals obtained from the Delsys EMG test system measurements as compared with the degree of muscle activation output by the OpenSim static optimization algorithm. In addition, joint angles and joint moments were compared to data obtained in other studies to verify the accuracy of the model.

2.4. Data Processing

The data were processed using Vicon Nexus (v.1.8.5, Vicon Motion system, Oxford, UK). The trajectory data of the marker points and ground reaction forces were filtered using 12 Hz and 60 Hz [8]. The filtering was performed by using a personalized MATLAB program (v2017a, The MathWorks Inc., Natick, Middlesex, MA, USA). Then, the trajectory and ground reaction force data were converted and processed using MATLAB (such that the c3d format files were converted to trc and mot formats usable for OpenSim input).

The weights of the marker points in the model were manually adjusted, and the model was scaled to match the anthropometric characteristics of the participants so that the root mean square error of the marker points and virtual marker points in the experiment was less than 0.02, and the maximum error was less than 0.04. Then, the joint angles and moments were calculated using the inverse kinematics and inverse dynamics algorithm tools in OpenSim. Static optimization was used to estimate the degree of muscle activation during exercise for the major muscle groups, including the rectus femoris, vastus medialis, vastus lateralis, vastus intermedius, biceps femoris, gastrocnemius, soleus, and tibialis anterior.

Hip and knee extension and metatarsal flexion of the ankle joint were defined as positive numbers. Joint power was calculated as the product of the joint moment and joint angular velocity (determined from kinematic data) [26]. Mechanical work was calculated as the joint power accumulated over time, with negative work values indicating the energy dissipation of the muscle due to eccentric contraction (e.g., the quadriceps muscle is elongated during knee flexion when the quadriceps is in an eccentric contraction state) [26,48–51]. Participants' power and mechanical work were normalized based on the sum of the participant's body weights and the weight of the barbell used.

In the present study, the work done by the limb was visualized with reference to the CLEW method proposed by Anahid et al. [35]. Absolute work was defined as the sum of positive and negative work done by the limbs (hip, knee, and ankle) throughout the squat cycle. Relative work was defined as the absolute value of the work done by each component as a percentage of the absolute total. The pie chart for the CLEW method in this study was created using the steps described in (Figure 1e).

2.5. Statistical Analysis

Normality and sphericity hypotheses were verified using Shapiro–Wilk's and Mauchly's tests, respectively. Means and standard deviations were calculated for kinematics, kinetics, and degree of muscle activation for each heel height condition. One-way repeated measures ANOVAs were performed using IBM SPSS Statistics (version 26, SPSS AG, Zurich, Switzerland), with Bonferroni's post hoc tests performed when necessary. If the sphericity assumption was violated, the degrees of freedom were adjusted using Greenhouse–Geisser correction. In addition, effect sizes were calculated using partial eta^2 (η_p^2), with the relative magnitude of any differences expressed as a standard criterion: small effect size = 0.01, medium effect size = 0.06, and large effect size = 0.14 [52]. Significance was set at $p < 0.05$.

3. Results

3.1. Model Validation

As shown in Figure 2, the muscle activation calculated using the OpenSim static optimization tool during barbell squatting is similar to the EMG signal activity recorded in the experiment, indicating that the data of the OpenSim model in this study is reliable [53,54]. The joint angle and moment obtained by the OpenSim inverse kinematics and inverse kinetics algorithm are similar to previous studies [8,23,47].

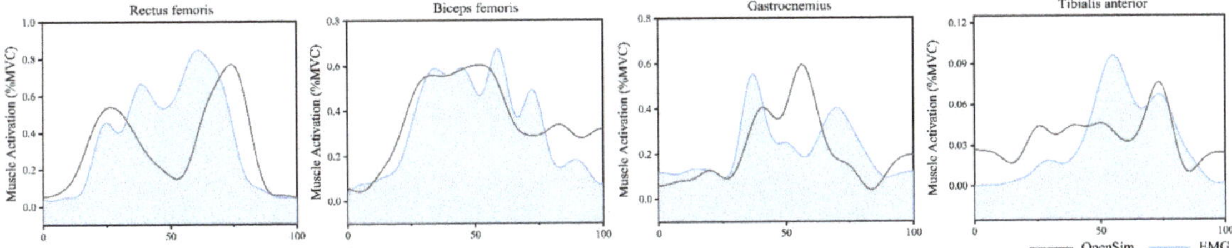

Figure 2. The results of rectus femoris, gastrocnemius, long biceps femoris, and anterior tibialis activation levels obtained using the EMG signal and the OpenSim optimization algorithm were more consistent.

3.2. Joint Angle, Joint Moment, and Joint Power

In Table 2, the results show no significant differences in joint angle, joint moment, and power of the hip joint regardless of the height of the insoles used ($p > 0.05$). Only male participants showed a significant increase in peak knee extension moment ($p < 0.001$, F = 19.434, $\eta_p^2 = 0.599$) and a significant increase in mean eccentric power ($p = 0.009$, F = 5.745, $\eta_p^2 = 0.307$) was observed when the heel was elevated using the 3 cm insole as compared with the 0 cm and 1.5 cm insoles. With increasing heel height, a significant main effect was observed both in the male group ($p = 0.001$, F = 10.870, $\eta_p^2 = 0.455$) and the female group ($p < 0.001$, F = 84.258, $\eta_p^2 = 0.849$), as well as the combined data from both genders ($p < 0.001$, F = 56.894, $\eta_p^2 = 0.662$); a significant main effect of maximum plantarflexion angle of the ankle was observed. In all participants ($p < 0.001$, F = 28.023, $\eta_p^2 = 0.491$) and in the female group ($p < 0.001$, F = 51.878, $\eta_p^2 = 0.776$), the maximum ankle dorsiflexion angle significantly decreased as the heel was elevated. In addition, the combined data from both genders showed greater ankle ROM when using 3 cm versus 1.5 cm insoles ($p = 0.047$, $F = 3.231$, $\eta_p^2 = 0.116$); when using 1.5 cm insoles, the ankle ROM of female subjects was significantly smaller than when using 0 cm insoles ($p = 0.011$, $F = 5.316$, $\eta_p^2 = 0.262$); when using 3 cm insoles, male subjects had significantly greater ankle ROM than when using 0 cm insoles ($p = 0.014$, $F = 5.087$, $\eta_p^2 = 0.281$).

Table 2. Joint angle, joint moment, and joint power.

Index		Male			Female			Average		
		0 cm	1.5 cm	3 cm	0 cm	1.5 cm	3 cm	0 cm	1.5 cm	3 cm
Hip	Max A (°)	−4.94 ± 9.37	−9.17 ± 4.53	−4.98 ± 13.44	−13.56 ± 5.11	−16.21 ± 4.67	−19.78 ± 4.23	−9.48 ± 8.54	−12.96 ± 5.79	−15.27 ± 10.67
	Min A (°)	−87.23 ± 9.45	−88.96 ± 9.27	−92.20 ± 12.67	−100.94 ± 8.21	−106.37 ± 7.24	−105.59 ± 13.34	−94.54 ± 11.15	−98.33 ± 11.97	−101.51 ± 14.51
	ROM (°)	82.29 ± 9.84	79.79 ± 7.60	87.23 ± 9.43	87.48 ± 8.44	90.16 ± 7.82	85.81 ± 12.89	85.06 ± 9.48	85.37 ± 9.29	86.24 ± 11.96
	Peak M (Nm/kg)	0.86 ± 0.12	0.81 ± 0.15	0.72 ± 0.02	0.68 ± 0.14	0.83 ± 0.20	0.79 ± 0.20	0.76 ± 0.16	0.82 ± 0.18	0.77 ± 0.17
	Peak CP (W/kg)	1.30 ± 0.34	1.28 ± 0.34	1.09 ± 0.24	0.90 ± 0.18	1.06 ± 0.22	1.00 ± 0.25	1.08 ± 0.33	1.17 ± 0.30	1.02 ± 0.25
	Peak EP (W/kg)	−0.86 ± 0.26	−0.98 ± 0.13	−0.97 ± 0.14	−0.75 ± 0.22	−0.93 ± 0.24	−0.97 ± 0.34	−0.80 ± 0.25	−0.94 ± 0.20	−0.97 ± 0.29
Knee	Max A (°)	−16.05 ± 7.04	−15.57 ± 3.33	−15.24 ± 5.97	−10.51 ± 7.25	−8.48 ± 4.79	−10.52 ± 5.82	−13.09 ± 7.67	−11.75 ± 5.48	−11.96 ± 6.25
	Min A (°)	−137.42 ± 3.97	−137.52 ± 3.92	−140.77 ± 4.60	−125.16 ± 10.51	−125.57 ± 5.84	−129.54 ± 7.80	−130.88 ± 10.18	−131.09 ± 7.81	−132.96 ± 8.68
	ROM (°)	121.38 ± 3.97	121.95 ± 3.26	125.53 ± 5.68	114.65 ± 16.89	117.09 ± 9.12	119.02 ± 11.81	117.79 ± 13.07	119.33 ± 7.45	121.00 ± 10.76
	Peak M (Nm/kg)	0.84 ± 0.08 [c]	0.89 ± 0.07 [c]	1.02 ± 0.10 [ab]	0.86 ± 0.10	0.83 ± 0.09	0.90 ± 0.10	0.85 ± 0.09	0.86 ± 0.09	0.93 ± 0.11
	Peak CP (W/kg)	1.07 ± 0.23	1.08 ± 0.16	1.24 ± 0.23	1.13 ± 0.34	1.03 ± 0.24	1.06 ± 0.18	1.10 ± 0.29	1.06 ± 0.21	1.11 ± 0.21
	Peak EP (W/kg)	−1.41 ± 0.32 [c]	−1.55 ± 0.34	−1.80 ± 0.38 [a]	−1.08 ± 0.21	−0.99 ± 0.22	−1.21 ± 0.28	−1.23 ± 0.31	−1.25 ± 0.39	−1.39 ± 0.41
Ankle	Max A (°)	−1.66 ± 3.83 [c]	0.29 ± 5.01	4.28 ± 3.01 [a]	6.47 ± 2.91 [bc]	9.33 ± 2.75 [ac]	16.08 ± 1.11 [ab]	2.67 ± 5.28 [c]	5.16 ± 6.00 [c]	12.49 ± 5.75 [ab]
	Min A (°)	−36.43 ± 2.65	−36.14 ± 1.90	−34.80 ± 1.24	−31.81 ± 4.37 [bc]	−26.12 ± 3.90 [ac]	−21.09 ± 2.82 [ab]	−33.97 ± 4.34 [bc]	−30.74 ± 5.90 [ac]	−25.27 ± 6.77 [ab]
	ROM (°)	34.77 ± 4.96 [c]	36.43 ± 3.50	39.08 ± 2.82 [a]	38.28 ± 2.75 [b]	35.45 ± 2.98 [a]	37.17 ± 2.70	36.64 ± 4.31	35.90 ± 3.27 [c]	37.75 ± 2.88 [b]
	Peak M (Nm/kg)	0.73 ± 0.11	0.68 ± 0.05	0.70 ± 0.07	0.47 ± 0.05	0.47 ± 0.08	0.49 ± 0.08	0.59 ± 0.16	0.57 ± 0.12	0.56 ± 0.13
	Peak CP (W/kg)	0.53 ± 0.14	0.56 ± 0.09	0.66 ± 0.16	0.35 ± 0.09	0.34 ± 0.09	0.35 ± 0.08	0.44 ± 0.14	0.44 ± 0.14	0.44 ± 0.18
	Peak EP (W/kg)	−0.35 ± 0.10	−0.36 ± 0.10	−0.33 ± 0.07	−0.22 ± 0.04	−0.24 ± 0.06	−0.22 ± 0.07	−0.28 ± 0.10	−0.29 ± 0.10	−0.25 ± 0.08

When participants are not gender specific: [a] significant differences with 0 cm, [b] significant differences with 1.5 cm, [c] significant differences with 3 cm (A, angle; M, moment; CP, concentric power; EP, eccentric power). The joint angle and joint moment data were obtained from OpenSim, and the joint power data were processed using the joint angle and joint moment data.

3.3. Joint Work

The proportion of relative work done by the joints during squatting is shown in Table 3 and Figure 3. The combined data from both genders showed that, when using the 3 cm insole, the percentage of positive work ($p < 0.001$, $F = 103.365$, $\eta_p^2 = 0.781$) and negative work ($p < 0.001$, $F = 17.008$, $\eta_p^2 = 0.370$) performed by the ankle joint were significantly greater than when using 0 cm and 1.5 cm insoles. In female subjects, the percentage of positive ($p < 0.001$, F = 67.250, $\eta_p^2 = 0.818$) and negative ($p = 0.001$, F = 14.721, $\eta_p^2 = 0.495$) work performed on the ankle joint was significantly greater with the 3 cm insole than with the 0 cm and 1.5 cm insoles. In male subjects, the negative work performed on the knee joint was significantly greater with the 3 cm heel height insole than with the 0 cm insole ($p = 0.040$, F = 3.032, $\eta_p^2 = 0.103$). In addition, a significant main effect was observed in male subjects with higher insoles for both positive ($p = 0.005$, F = 6.699, $\eta_p^2 = 0.34$) and negative ($p = 0.006$, F = 6.281, $\eta_p^2 = 0.344$) work done by the hip joint and the percentage of positive work done by the ankle joint ($p < 0.001$, F = 44.394, $\eta_p^2 = 0.773$) both showed a significant main effect, with the percentage of work done by the hip decreasing and the ankle increasing with greater heel height.

Table 3. The relative proportion of work done by joints.

Index		Male			Female			Average		
		0 cm	1.5 cm	3 cm	0 cm	1.5 cm	3 cm	0 cm	1.5 cm	3 cm
Hip	NWR	0.32 ± 0.06 [c]	0.31 ± 0.07 [c]	0.25 ± 0.04 [ab]	0.36 ± 0.08	0.44 ± 0.09	0.40 ± 0.12	0.34 ± 0.07	0.38 ± 0.10	0.35 ± 0.12
	PWR	0.38 ± 0.04 [c]	0.36 ± 0.05	0.32 ± 0.04 [a]	0.37 ± 0.06	0.44 ± 0.08	0.41 ± 0.10	0.38 ± 0.05	0.40 ± 0.08	0.38 ± 0.09
Knee	NWR	0.56 ± 0.08 [c]	0.57 ± 0.10	0.63 ± 0.06 [a]	0.54 ± 0.08	0.46 ± 0.10	0.52 ± 0.13	0.55 ± 0.08	0.51 ± 0.11	0.55 ± 0.13
	PWR	0.48 ± 0.06	0.50 ± 0.06	0.52 ± 0.05	0.50 ± 0.05	0.45 ± 0.08	0.48 ± 0.09	0.49 ± 0.06	0.47 ± 0.08	0.49 ± 0.08
Ankle	NWR	0.12 ± 0.04	0.12 ± 0.03	0.16 ± 0.05	0.09 ± 0.01 [c]	0.09 ± 0.02 [c]	0.15 ± 0.05 [ab]	0.10 ± 0.03 [c]	0.10 ± 0.03 [c]	0.15 ± 0.05 [ab]
	PWR	0.14 ± 0.03 [c]	0.15 ± 0.02 [c]	0.24 ± 0.03 [ab]	0.13 ± 0.02 [bc]	0.11 ± 0.01 [ac]	0.20 ± 0.04 [ab]	0.13 ± 0.03 [c]	0.13 ± 0.02 [c]	0.21 ± 0.04 [ab]

When participants are not gender specific: [a] significant differences with 0 cm, [b] significant differences with 1.5 cm, [c] significant differences with 3 cm (NWR, negative work ratio and PWR, positive work ratio). The percentage of work done by the joints was obtained by processing the joint power data.

3.4. Muscle Activation

The average degree of muscle activation during squatting is shown in Figure 4. In male subjects, there was a significant main effect of mean muscle activation in the vastus lateralis ($p < 0.001$, F = 10.870, $\eta_p^2 = 0.364$) and anterior tibialis ($p < 0.001$, F = 11.854, $\eta_p^2 = 0.384$) during deep squatting with different insole heights. Post hoc pairwise comparisons showed that the mean muscle activation of the vastus lateralis was significantly greater in male subjects with the 3 cm insole than with the 0 cm and 1.5 cm insoles. The average muscle activation of the anterior tibialis muscle decreased significantly with increasing insole height.

In female subjects, there was a significant main effect of mean muscle activation in the short head of the biceps femoris ($p = 0.003$, F = 6.651, $\eta_p^2 = 0.259$), anterior tibialis ($p = 0.01$, F = 5.183, $\eta_p^2 = 0.214$), and medial gastrocnemius ($p = 0.018$, F = 4.508, $\eta_p^2 = 0.192$) when using different insole heights. Post hoc pairwise comparisons showed that the mean muscle activation of the short head of the biceps femoris was significantly greater in female subjects during the barbell squat when using a 3 cm insole than when using a 0 cm insole. When using a 1.5 cm insole, the average muscle activation of the anterior tibialis was significantly less than when using a 0 cm insole. With an increase in the heel height, the average muscle activation of the medial gastrocnemius increased; the average muscle activation of the medial gastrocnemius increased when using a 1.5 cm insole. The combined data from both genders showed that the mean muscle activation of the anterior tibialis showed a significant main effect ($p = 0.01$, F = 6.082, $\eta_p^2 = 0.403$) when using different height insoles, and the mean muscle activation of the medial gastrocnemius increased when using 1.5 cm and 3 cm insoles. The mean muscle activation of the anterior tibialis was significantly greater with 1.5 cm and 3 cm insoles than with 0 cm insoles.

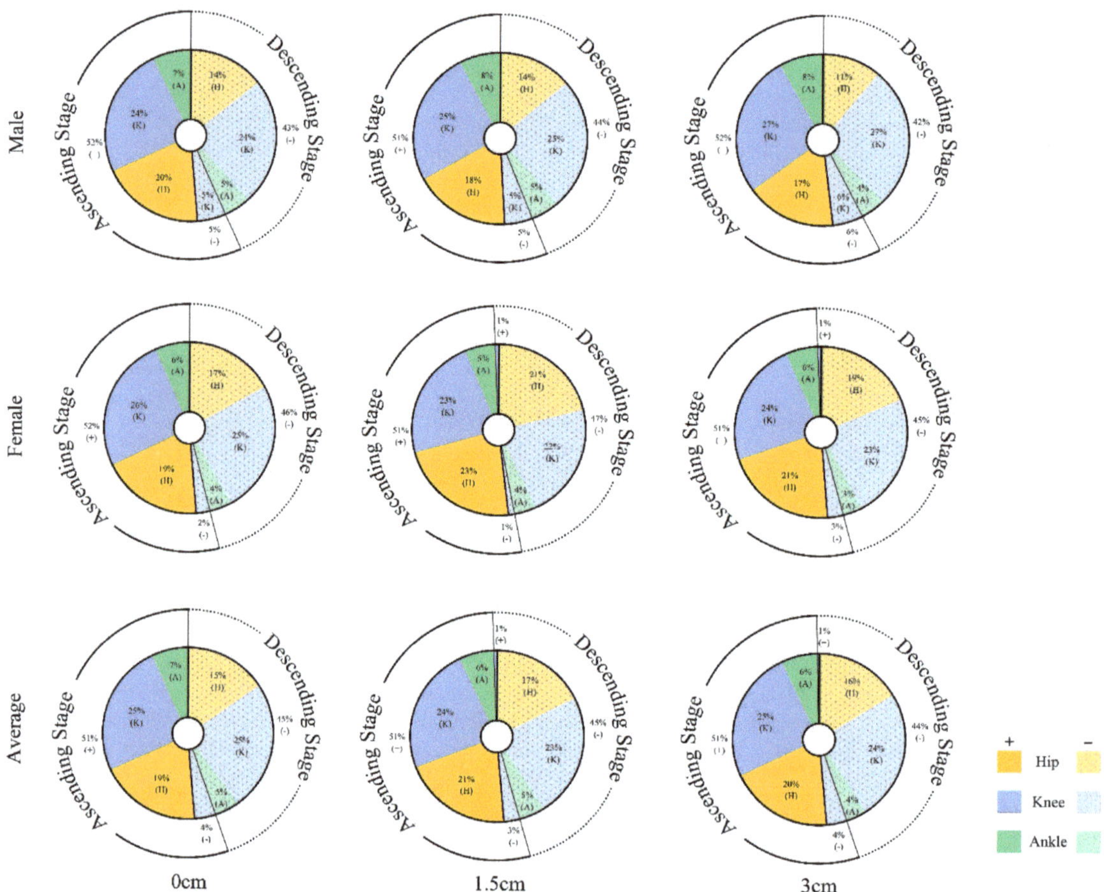

Figure 3. The percentage of work done by the hip, knee, and ankle joints during the descent and ascent of the barbell squat for different genders and heel heights. The percentage of work done by the joints was obtained by processing the joint power data.

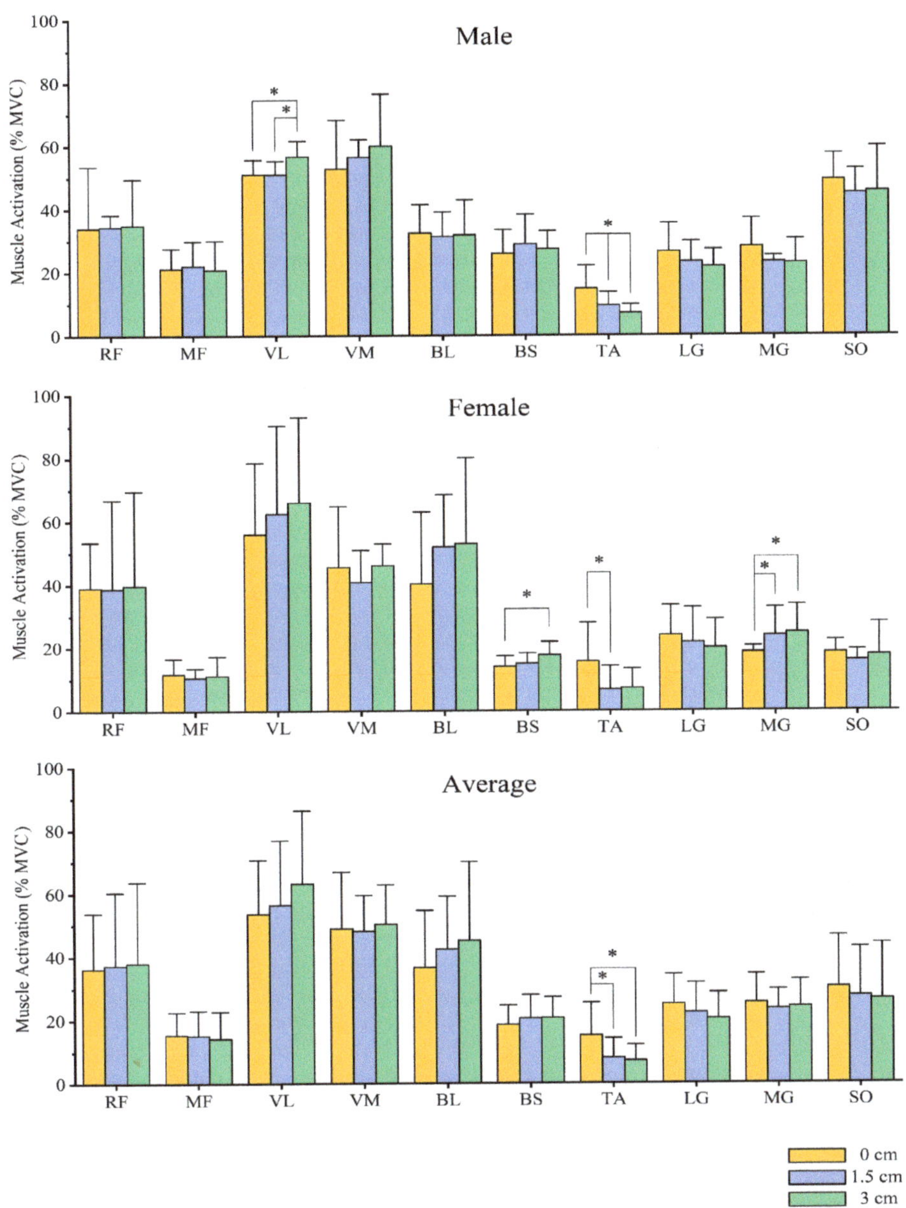

Figure 4. Mean standardized muscle activation during the barbell squat in males, females, and all participants. * Significant differences between the two groups. (RF, rectus femoris; MF, middle femoral; VL, vastus lateralis; VM, vastus medialis; BL, biceps femoris long head; BS, biceps femoris short head; TA, tibialis anterior; LG, lateral gastrocnemius; MG, medial gastrocnemius; SO, soleus). The OpenSim static optimization algorithm obtained the muscle activation data.

4. Discussion

Barbell squats are commonly used for strength training and rehabilitation [1–15]. Heel elevation during a barbell squat is thought to improve deep squat movement patterns.

However, current research on the effects of heel elevation on the deep squat has focused on kinematics, with little evidence on aspects such as kinetics and muscle activity. Therefore, in this study, we used the OpenSim musculoskeletal model customized for the deep squat to compare the effects of different heels on joint angles, joint moments, joint power, joint work done, and the degree of muscle activation during the barbell deep squat.

4.1. Joint Angle, Joint Moment, and Power

Participants' ankle angles were directly affected during the barbell squat when heel height was changed. In the present study, it was observed that as heel height increased, participants exhibited progressively greater ankle plantarflexion and progressively less ankle dorsiflexion during the barbell squat, which was consistent with our hypothesis. When the heel is raised, the ankle is forced into a higher degree of plantarflexion, decreasing the ankle dorsiflexion angle [12,18,23,55,56], which may indirectly affect the knee or trunk angle [11]. In contrast, Charlton et al. showed that participants' ankle dorsiflexion angles did not change with heel height [57]. This difference in results may be due to the experimental design. In the study by Charlton et al., participants used a uniform weight of 20 kg, which was well below the participants' regular weights in training. While smaller weights may allow participants to squat with minimal fatigue or technical bias [57], higher weights may have other effects on movement [15], therefore, the results by Charlton et al. might not fully explain the effect of heel height. In conclusion, individuals with low ankle ROM often find it difficult to perform large dorsiflexion movements, therefore, there might be potential benefits to using heel elevation insoles during the barbell squat.

Similar to the ankle, the kinematics of the knee joint are very easily observed during the barbell deep squat. Raising the heel has been observed to increase knee ROM and extension moment in many previous studies [18,20,24,57], similar to the results obtained in this study. However, in this study, raising the heel height was observed to increase the moment and mean eccentric power only in male participants, but did not cause a significant change in knee mobility, which was partially consistent with our hypothesis. The differences in experimental design may explain the conflicting results in knee angles, such as whether the control group wore the same style of footwear, whether the depth of the squat had been maximized, and individual differences in squatting habits. In this study, participants were all wearing uniform footwear to control these factors and received instruction from the same coach not associated with the study. In a recent study [17], Monteiro et al. observed that the ROM of the knee joint increased with heel elevation during the barbell squat. Still, the difference in knee mobility was only statistically significant when the heel was elevated by 5 cm. Although the knee kinematics were not statistically significant in this study, there was still a tendency for peak knee flexion and mobility to be greater with an increase in heel height in all three cases. Therefore, elevating the heel may still have a potential effect on knee joint angle during the barbell squat, which is consistent with the findings of Monteiro et al. In addition, however, a significant main effect of knee moment was observed only in male participants, and there was a gradually increasing trend in knee moment and in power as the heel height increased. The results of biomechanical testing usually need to be analyzed in conjunction with the purpose of the intervention. This study suggests that heel elevation may increase knee extension moment and power during deep squats, implying that the knee joint might be subjected to greater loading. Still, this change in movement pattern is significant for individuals seeking strength, muscle hypertrophy, and rehabilitation; therefore, a moderate increase in heel height may enhance the knee joint training effect. The differences in experimental design may explain the conflicting results in knee angles, such as whether the control group wore the same style of footwear, whether the depth of the squat had been maximized, and individual differences in squatting habits. In this study, participants were all wearing uniform footwear to control these factors and received instruction from the same trainer not associated with the study. In a recent study [17], Monteiro et al. observed that knee ROM increased with heel elevation during the barbell squat. Still, a significant difference was only shown when the heel was elevated by 5 cm. Similar to the findings of Monteiro et al., although knee kinematics did not show significant differences in our study, there was still a tendency for peak knee flexion and mobility to be greater with an increase in heel height. Therefore, heel elevation may still have a potential effect on knee angulation. Furthermore, although a significant main effect of knee moment was only observed in male participants, all participants also showed a gradual increase in knee moment and in power with increasing heel height. In conclusion, heel elevation during the deep squat may increase knee extension moment and power, but it also means that the knee joint may be subjected to greater loading. Nevertheless, this change in movement pattern is important for those seeking strength, muscle hypertrophy, and rehabilitation; therefore, a moderate increase in heel height may enhance the knee joint training effect.

In this study, no statistically significant differences in hip joint angle, joint moment, or power were found regardless of the height of the insole used by the participants in the barbell squat. Although it is widely believed that elevating the heel during the barbell squat might result in a more upright trunk to prevent back injury, the current findings suggest that heel height only has little effect on hip flexion angle [20,23,24,58] and no statistically significant effect on hip and lower back muscle activation was found [58]. Notably, although raising the heel had little effect on the hip flexion angle during the barbell squat, the tilt angle of the trunk decreased with heel elevation in the study by Monteiro et al. [17]. This may result in a biomechanical transfer from the hip joint to the knee joint, contributing to the increased knee moment.

4.2. Joint Work

A comparative analysis of the proportion of lower extremity joints doing work showed that the proportion of hip joint work decreased with increasing heel height. In contrast, the proportion of ankle joint work increased gradually. Combined with the results of other indices, it can be hypothesized that when heel height is increased, male participants prefer to use the knee joint rather than the hip joint to lift the barbell. In contrast, the ankle joint needs to improve plantarflexion performance to enhance trunk stability. Although an increase in ankle work ratio was also observed in female participants, the knee work ratio did not change significantly with the increased heel height. In contrast, the hip work ratio increased with a 1.5 cm heel, which may be influenced by the physiological differences between gender and the level of training. To the best of our knowledge, this is the first attempt to analyze the effect of heel height on the lower extremity joint work ratio. Therefore, the current evidence could not draw definitive conclusions about the specific effects of heel height on lower extremity joint work during the barbell squat, and more evidence would be helpful.

4.3. Muscle Activation

This study suggests that heel elevation may increase activation of the knee extensor and the metatarsal flexor muscle group, similar to the results of previous studies [40,59]. In a study by Christopher et al. [40], although the results were not statistically significant, there was a slight increase in muscle activation of the lateral vastus and medial vastus muscles with a raised heel, and the average muscle activation of the gastrocnemius was significantly higher when raising the heel [40,59], which was similar to our results. This might be the result of the raised heel causing the movement of the body's center of mass and the center of pressure, which forces participants to activate the metatarsal flexor muscle group to control balance and resist the body's center of mass forward during the squat [11,40]. In several previous studies, differences in muscle activation only reached statistically significant levels when the heel was raised high enough [40,59]. In this study, the mean muscle activation of the dorsiflexion and plantarflexion muscle groups was significantly altered when the heel was raised by 1.5 cm, and the mean activation of the vastus lateralis and biceps femoris short head was increased considerably when raised by 3 cm. These results are essential for both training and rehabilitation. For athletes who want to improve the strength of the knee joint muscles, heel elevation can be used to increase the activation of the knee muscles to improve training results, and the activation of the ankle flexors and extensors can be an important reference for developing ankle stability training.

4.4. Limitations

This study still has some limitations. The main results are as follows: (1) The study of kinematics and kinetics are only considered in the sagittal plane. (2) Although the participants were all enthusiasts who had undergone strength training for more than one year, the maximum weight of barbell squats in males was no more than two times their body weight, and that in women was about one times their body weight. However, it is impossible to know whether the results of this study apply to higher level lifters.

5. Conclusions

This study showed that although some of the statistical differences were minor, raising the heel might still affect the joint ROM, joint moments, and muscle activation in the lower extremity and redistribute the biomechanical loads between the hip, knee, and ankle joints during barbell squats. These minor changes might affect musculoskeletal adaptations, essential for training and rehabilitation. However, the degree to which different heel heights affect people of different genders appears to be different, but in all participants, raising the heel reduces the dorsiflexion angle and muscle activation of the dorsiflexors, and increases the proportion of mechanical work performed by the ankle joint. Therefore, changes in heel height might have a potential effect on the technical movements of the barbell squat. Furthermore, regardless of the data results, athletes still report positive effects from elevated heel height, which might be psychological. Nevertheless, the results might vary between individuals with different training experiences and physical characteristics, and therefore, carefully choosing the shoes is necessary. In conclusion, for those with lower back and ankle injuries or mobility limitations, or those who wish to improve knee engagement during the barbell squat, we still recommend elevating the heel during the barbell squat to a height of no less than 1.5 cm.

Author Contributions: Conceptualization, Z.L., X.L. and Y.G.; methodology, X.L., R.X., M.L. and I.B.; software, Z.L., X.L., Y.S. and Y.G.; validation, X.L., R.X. and I.B.; investigation, Y.G., M.L. and I.B.; writing—original draft preparation, Z.L. and M.L.; writing—review and editing, R.X., I.B. and Y.G. All authors have read and agreed to the published version of the manuscript.

Funding: This study was sponsored by the National Natural Science Foundation of China (No. 81772423), Key Project of the National Social Science Foundation of China (19ZDA352), Key R&D Program of Zhejiang Province China (2021C03130), and Ningbo Public Welfare Science and Technology Plan Project (No.2019C50095).

Institutional Review Board Statement: This study was conducted according to the guidelines of the Declaration of Helsinki and approved by the Ethics Committee of Ningbo University (RAGH202101150322.6).

Informed Consent Statement: Informed consent was obtained from all participants involved in the study.

Data Availability Statement: The data that support the findings of this study are available on reasonable request from the corresponding author. The data are not publicly available due to privacy or ethical restrictions.

Conflicts of Interest: The authors declare no conflict of interest.

References

1. Comfort, P.; Kasim, P. Optimizing squat technique. *Strength Cond. J.* **2007**, *29*, 10. [CrossRef]
2. Corradi, E.F.; Lanza, M.B.; Lacerda, L.T.; Andrushko, J.W.; Martins-Costa, H.C.; Diniz, R.C.; Lima, F.V.; Chagas, M.H. Acute physiological responses with varying load or time under tension during a squat exercise: A randomized cross-over design. *J. Sci. Med. Sport* **2021**, *24*, 171–176. [CrossRef] [PubMed]
3. Cotter, J.A.; Chaudhari, A.M.; Jamison, S.T.; Devor, S.T. Knee joint kinetics in relation to commonly prescribed squat loads and depths. *J. Strength Cond. Res./Natl. Strength Cond. Assoc.* **2013**, *27*, 1765. [CrossRef] [PubMed]

4. Shi, Z.; Sun, D. Conflict between Weightlifting and Health? The Importance of Injury Prevention and Technology Assistance. *Phys. Act. Health* **2022**, *6*, 1–4. [CrossRef]

5. Fry, A.C.; Smith, J.C.; Schilling, B.K. Effect of knee position on hip and knee torques during the barbell squat. *J. Strength Cond. Res.* **2003**, *17*, 629–633. [PubMed]

6. Lee, S.P.; Gillis, C.B.; Ibarra, J.J.; Oldroyd, D.F.; Zane, R.S. Heel-raised foot posture does not affect trunk and lower extremity biomechanics during a barbell back squat in recreational weight lifters. *J. Strength Cond. Res.* **2019**, *33*, 606–614. [CrossRef]

7. List, R.; Gülay, T.; Stoop, M.; Lorenzetti, S. Kinematics of the trunk and the lower extremities during restricted and unrestricted squats. *J. Strength Cond. Res.* **2013**, *27*, 1529–1538. [CrossRef]

8. Yichen, L.; Qichang, M.; Peng, H.-T.; Jianshe, L.; Chen, W.; Yaodong, G. A comparative study on loadings of the lower extremity during deep squat in Asian and Caucasian individuals via OpenSim musculoskeletal modelling. *BioMed Res. Int.* **2020**, *2020*, 7531719.

9. Lynn, S.K.; Noffal, G.J. Lower extremity biomechanics during a regular and counterbalanced squat. *J. Strength Cond. Res.* **2012**, *26*, 2417–2425. [CrossRef]

10. McKean, M.R.; Dunn, P.K.; Burkett, B.J. Quantifying the movement and the influence of load in the back squat exercise. *J. Strength Cond. Res.* **2010**, *24*, 1671–1679. [CrossRef]

11. Pangan, A.M.; Leineweber, M. Footwear and elevated heel influence on barbell back squat: A review. *J. Biomech. Eng.* **2021**, *143*, 090801. [CrossRef] [PubMed]

12. Schoenfeld, B.J. Squatting kinematics and kinetics and their application to exercise performance. *J. Strength Cond.* **2010**, *24*, 3497–3506. [CrossRef] [PubMed]

13. Swinton, P.A.; Lloyd, R.; Keogh, J.W.; Agouris, I.; Stewart, A.D. A biomechanical comparison of the traditional squat, powerlifting squat, and box squat. *J. Strength Cond. Res.* **2012**, *26*, 1805–1816. [CrossRef] [PubMed]

14. Walsh, J.C.; Quinlan, J.F.; Stapleton, R.; FitzPatrick, D.P.; McCormack, D. Three-dimensional motion analysis of the lumbar spine during "free squat" weight lift training. *Am. J. Sports Med.* **2007**, *35*, 927–932. [CrossRef] [PubMed]

15. Whitting, J.W.; Meir, R.A.; Crowley-McHattan, Z.J.; Holding, R.C. Influence of footwear type on barbell back squat using 50, 70, and 90% of one repetition maximum: A biomechanical analysis. *J. Strength Onditioning Res.* **2016**, *30*, 1085–1092. [CrossRef] [PubMed]

16. Kongsgaard, M.; Aagaard, P.; Roikjaer, S.; Olsen, D.; Jensen, M.; Langberg, H.; Magnusson, S. Decline eccentric squats increases patellar tendon loading compared to standard eccentric squats. *Clin. Biomech.* **2006**, *21*, 748–754. [CrossRef]

17. Monteiro, P.; Marcori, A.J.; Nascimento, V.; Guimarães, A.; Okazaki, V.H.A. Comparing the kinematics of back squats performed with different heel elevations. *Hum. Mov.* **2022**, *23*, 97–103. [CrossRef]

18. Sato, K.; Fortenbaugh, D.; Hydock, D.S. Kinematic changes using weightlifting shoes on barbell back squat. *J. Strength Cond. Res.* **2012**, *26*, 28–33. [CrossRef]

19. Schermoly, T.P.; Hough, I.G.; Senchina, D.S. The effects of footwear on force production during barbell back squats. *J. Undergrad. Kinesiol. Res.* **2015**, *10*, 42–51.

20. Southwell, D.J.; Petersen, S.A.; Beach, T.A.; Graham, R.B. The effects of squatting footwear on three-dimensional lower limb and spine kinetics. *J. Electromyogr. Kinesiol.* **2016**, *31*, 111–118. [CrossRef]

21. Zobeck, C.J.; Senchina, D.S. The Effects of Footwear on Force Production during barbell back squats part II: Standardizing squat depth. *Int. J. Res. Exerc. Physiol.* **2020**, *15*, 24–34.

22. Legg, H.S.; Glaister, M.; Cleather, D.J.; Goodwin, J.E. The effect of weightlifting shoes on the kinetics and kinematics of the back squat. *J. Sports Sci.* **2017**, *35*, 508–515. [CrossRef] [PubMed]

23. Sinclair, J.; McCarthy, D.; Bentley, I.; Hurst, H.T.; Atkins, S. The influence of different footwear on 3-D kinematics and muscle activation during the barbell back squat in males. *Eur. J. Sport Sci.* **2015**, *15*, 583–590. [CrossRef]

24. Sato, K.; Fortenbaugh, D.; Hydock, D.S.; Heise, G.D. Comparison of back squat kinematics between barefoot and shoe conditions. *Int. J. Sports Sci. Coach.* **2013**, *8*, 571–578. [CrossRef]

25. Sayers, M.G.; Bachem, C.; Schütz, P.; Taylor, W.R.; List, R.; Lorenzetti, S.; Nasab, S.H. The effect of elevating the heels on spinal kinematics and kinetics during the back squat in trained and novice weight trainers. *J. Sports Sci.* **2020**, *38*, 1000–1008. [CrossRef]

26. Jandacka, D.; Uchytil, J.; Farana, R.; Zahradnik, D.; Hamill, J. Lower extremity power during the squat jump with various barbell loads. *Sports Biomech.* **2014**, *13*, 75–86. [CrossRef] [PubMed]

27. Kaneko, M. Training effect of different loads on thd force-velocity relationship and mechanical power output in human muscle. *Scand. J. Sports Sci.* **1983**, *5*, 50–55.

28. Wilkie, D. The relation between force and velocity in human muscle. *J. Physiol.* **1949**, *110*, 249. [CrossRef]

29. Wakeling, J.; Blake, O.; Chan, H. Muscle coordination is key to the power output and mechanical efficiency of limb movements. *J. Exp. Biol.* **2010**, *213*, 487–492. [CrossRef]

30. Teixeira-Salmela, L.F.; Nadeau, S.; Milot, M.-H.; Gravel, D.; Requião, L.F. Effects of cadence on energy generation and absorption at lower extremity joints during gait. *Clin. Biomech.* **2008**, *23*, 769–778. [CrossRef]

31. Sawicki, G.S.; Lewis, C.L.; Ferris, D.P. It pays to have a spring in your step. *Exerc. Sport Sci. Rev.* **2009**, *37*, 130. [CrossRef] [PubMed]

32. Cofré, L.E.; Lythgo, N.; Morgan, D.; Galea, M.P. Aging modifies joint power and work when gait speeds are matched. *Gait Posture* **2011**, *33*, 484–489. [CrossRef] [PubMed]

33. Chen, I.; Kuo, K.; Andriacchi, T. The influence of walking speed on mechanical joint power during gait. *Gait Posture* **1997**, *6*, 171–176. [CrossRef]

34. Silverman, A.K.; Fey, N.P.; Portillo, A.; Walden, J.G.; Bosker, G.; Neptune, R.R. Compensatory mechanisms in below-knee amputee gait in response to increasing steady-state walking speeds. *Gait Posture* **2008**, *28*, 602–609. [CrossRef] [PubMed]

35. Ebrahimi, A.; Goldberg, S.R.; Wilken, J.M.; Stanhope, S.J. Constituent Lower Extremity Work (CLEW) approach: A novel tool to visualize joint and segment work. *Gait Posture* **2017**, *56*, 49–53. [CrossRef] [PubMed]

36. Zelik, K.E.; Honert, E.C. Ankle and foot power in gait analysis: Implications for science, technology and clinical assessment. *J. Biomech.* **2018**, *75*, 1–12. [CrossRef]

37. Zellers, J.A.; Marmon, A.R.; Ebrahimi, A.; Grävare Silbernagel, K. Lower extremity work along with triceps surae structure and activation is altered with jumping after Achilles tendon repair. *J. Orthop. Res.* **2019**, *37*, 933–941. [CrossRef] [PubMed]

38. Sinclair, J.; Taylor, P.; Currigan, G.; Hobbs, S. The test-retest reliability of three different hip joint centre location techniques. *Mov. Sport Sci.-Sci. Mot.* **2014**, *83*, 31–39. [CrossRef]

39. Anbarian, M.; Rajabian, F.; Ghasemi, M.H.; Heidari Moghaddam, R. The effect of the heel wedges on the electromyography activities of the selected lower back muscles during load lifting. *Iran. J. Ergon.* **2017**, *5*, 12–21.

40. Johnston, C.; Hunt-Murray, C.; Hsieh, C. Effect of heel heights on lower extremity muscle activation for back-squat performance. *ISBS Proc. Arch.* **2017**, *35*, 176.

41. Faul, F.; Erdfelder, E.; Lang, A.-G.; Buchner, A. G* Power 3: A flexible statistical power analysis program for the social, behavioral, and biomedical sciences. *Behav. Res. Methods* **2007**, *39*, 175–191. [CrossRef] [PubMed]

42. Stock, M.S.; Beck, T.W.; DeFreitas, J.M.; Dillon, M.A. Test–retest reliability of barbell velocity during the free-weight bench-press exercise. *J. Strength Cond. Res.* **2011**, *25*, 171–177. [CrossRef] [PubMed]

43. Xiang, L.; Mei, Q.; Wang, A.; Shim, V.; Fernandez, J.; Gu, Y. Evaluating function in the hallux valgus foot following a 12-week minimalist footwear intervention: A pilot computational analysis. *J Biomech.* **2022**, *132*, 110941. [CrossRef] [PubMed]

44. He, Y.; Fekete, G. The Effect of Cryotherapy on Balance Recovery at Different Moments after Lower Extremity Muscle Fatigue. *Phys. Act. Health* **2021**, *5*, 255–270. [CrossRef]

45. Pereira, M.I.R.; Gomes, P.S.C. Muscular strength and endurance tests: Reliability and prediction of one repetition maximum-Review and new evidences. *Rev. Bras. Med. Esporte* **2003**, *9*, 325–335. [CrossRef]

46. Catelli, D.S.; Wesseling, M.; Jonkers, I.; Lamontagne, M. A musculoskeletal model customized for squatting task. *Comput. Methods Biomech. Biomed. Eng.* **2019**, *22*, 21–24. [CrossRef]

47. Xin, L.; Adrien, N.; Baker, J.S.; Qichang, M.; Yaodong, G. Novice female exercisers exhibited different biomechanical loading profiles during full-squat and half-squat practice. *Biology* **2021**, *10*, 1184.

48. Decker, M.J.; Torry, M.R.; Wyland, D.J.; Sterett, W.I.; Steadman, J.R. Gender differences in lower extremity kinematics, kinetics and energy absorption during landing. *Clin. Biomech.* **2003**, *18*, 662–669. [CrossRef]

49. Devita, P.; Skelly, W.A. Effect of landing stiffness on joint kinetics and energetics in the lower extremity. *Med. Sci. Sports Exerc.* **1992**, *24*, 108–115. [CrossRef]

50. Winter, D.A. *Biomechanics and Motor Control of Human Movement*; John Wiley & Sons: Hoboken, NJ, USA, 2009.

51. Kulas, A.; Zalewski, P.; Hortobagyi, T.; DeVita, P. Effects of added trunk load and corresponding trunk position adaptations on lower extremity biomechanics during drop-landings. *J. Biomech.* **2008**, *41*, 180–185. [CrossRef]

52. Cohen, J. A power primer. *Psychol. Bull.* **1992**, *112*, 155. [CrossRef] [PubMed]

53. Hamner, S.R.; Delp, S.L. Muscle contributions to fore-aft and vertical body mass center accelerations over a range of running speeds. *J. Biomech.* **2013**, *46*, 780–787. [CrossRef] [PubMed]

54. Rajagopal, A.; Dembia, C.L.; DeMers, M.S.; Delp, D.D.; Hicks, J.L.; Delp, S.L. Full-body musculoskeletal model for muscle-driven simulation of human gait. *IEEE Trans. Biomed. Eng.* **2016**, *63*, 2068–2079. [CrossRef] [PubMed]

55. Xu, D.; Quan, W.; Zhou, H.; Sun, D.; Baker, J.S.; Gu, Y. Explaining the differences of gait patterns between high and low-mileage runners with machine learning. *Sci Rep.* **2022**, *12*, 2981. [CrossRef] [PubMed]

56. Dempster, J.; Dutheil, F.; Ugbolue, U.C. The Prevalence of Lower Extremity Injuries in Running and Associated Risk Factors: A Systematic Review. *Phys. Act. Health* **2021**, *5*, 133–145. [CrossRef]

57. Charlton, J.M.; Hammond, C.A.; Cochrane, C.K.; Hatfield, G.L.; Hunt, M.A. The effects of a heel wedge on hip, pelvis and trunk biomechanics during squatting in resistance trained individuals. *The J. Strength Cond. Res.* **2017**, *31*, 1678–1687. [CrossRef]

58. Ibarra, J.; Oldroyd, D.; Zane, R. Heel-Raised Foot Posture and Weightlifting Shoes Do Not Affect Trunk and LowerExtremity Biomechanics during a Barbell Back Squat. Ph.D. Thesis, University of Nevada, Las Vegas, NV, USA, 2016.

59. Sriwarno, A.B.; Shimomura, Y.; Iwanaga, K.; Katsuura, T. The effects of heel elevation on postural adjustment and activity of lower-extremity muscles during deep squatting-to-standing movement in normal subjects. *J. Phys. Ther. Sci.* **2008**, *20*, 31–38. [CrossRef]

 *bioengineering*

Article

The Bionic High-Cushioning Midsole of Shoes Inspired by Functional Characteristics of Ostrich Foot

Rui Zhang [1,*], Liangliang Zhao [1], Qingrui Kong [1], Guolong Yu [1], Haibin Yu [2,*], Jing Li [2] and Wei-Hsun Tai [2]

1 Key Laboratory of Bionic Engineering, Ministry of Education, Jilin University, Changchun 130022, China
2 School of Physical Education, Quanzhou Normal University, Quanzhou 362000, China
* Correspondence: zhangrui@jlu.edu.cn (R.Z.); yhb@qztc.edu.cn (H.Y.)

Abstract: The sole is a key component of the interaction between foot and ground in daily activities, and its cushioning performance plays a crucial role in protecting the joints of lower limbs from impact injuries. Based on the excellent cushioning performance of the ostrich foot and inspired by the structure and material assembly features of the ostrich foot's metatarsophalangeal skeletal–tendon and the ostrich toe pad–fascia, a functional bionic cushioning unit for the midsole (including the forefoot and heel) area of athletic shoes was designed using engineering bionic technology. The bionic cushioning unit was then processed based on the bionic design model, and the shoe soles were tested with six impact energies ranging from 3.3 J to 11.6 J for a drop hammer impact and compared with the conventional control sole of the same size. The results indicated that the bionic forefoot area absorbed 9.83–34.95% more impact and 10.65–43.84% more energy than the conventional control forefoot area, while the bionic heel area absorbed 26.34–44.29% more impact and 28.1–51.29% more energy than the conventional control heel area when the controlled impact energy varied from 3.3 J to 11.6 J. The cushioning performance of the bionic cushioning sole was generally better than that of the conventional control sole, and the cushioning and energy-absorption performances of the heel bionic cushioning unit were better than those of the forefoot bionic cushioning unit. This study provides innovative reference and research ideas for the design and development of sports shoes with good cushioning performance.

Keywords: ostrich toe pad; metatarsophalangeal joint; high cushioning; midsole of soles; bionic design

Citation: Zhang, R.; Zhao, L.; Kong, Q.; Yu, G.; Yu, H.; Li, J.; Tai, W.-H. The Bionic High-Cushioning Midsole of Shoes Inspired by Functional Characteristics of Ostrich Foot. *Bioengineering* 2023, 10, 1. https://doi.org/10.3390/bioengineering10010001

Academic Editor: Stuart Goodman

Received: 17 November 2022
Revised: 13 December 2022
Accepted: 16 December 2022
Published: 20 December 2022

1. Introduction

As people attach greater and greater importance to exercise and health, their criterion for selecting footwear has shifted from aesthetic to functional [1]. Cushioning performance is one of the basic functional indicators of shoe soles, which not only plays a crucial role in the design and development of sports shoes, but also is an important factor in the evaluation of wearing comfort [2,3]. The sole of a shoe consists of insole, midsole, and outsole, among which midsole is the core of its cushioning technology. How to reasonably configure the material and structure of the midsole has become a major concern in the cushioning technology of shoe soles [4–6]. In fact, modern midsoles are generally made of foam-like materials processed to produce resistance at the foot–ground interface through the compression and elasticity of the material, consuming the mechanical energy generated by collision, thus playing a role in reducing impact load [7]. Generally speaking, the softer the material and the thicker the midsole, the better the cushioning performance, but accompanied by higher risks of injuries and lower stability [8]. Relying solely on material flexibility is bound to be contradictory in terms of cushioning performance and stability, which makes it an effective means to deal with the limited midsole thickness space, especially the thinner midsole space of the front foot from the perspective of structural and material coupling [9–11]. At the moment, major manufacturers primarily focus on aesthetics while ignoring functionality, and existing research focuses on the heel region, with targeted materials and structural configurations in the forefoot region and heel region, while there is little research on the improvement of the cushioning performance of the sole [12,13].

Improving the cushioning performance of sports shoes based on the engineering bionic principle has gradually attracted attention and begun to be applied in the current structural cushioning technology of sports shoes [14]. Traditional sports shoes limit the release of energy stored in the Achilles tendon and calf muscle during lift-off, resulting in a long stride, which is the cause of excessive braking force and impact force [15–18]. The African ostrich is known for its heavy-duty, high-speed, and long-lasting athletic characteristics. Adult African ostriches can sustain a speed of about 55 km/h for more than 30 min while supporting their own weight of up to 150 kg [19]. Existing research on the cushioning characteristics of the ostrich foot mainly focuses on the foot pad and metatarsophalangeal joint. S.A.A. El-Gendy et al. found through anatomical studies that the ostrich foot pad was able to effectively absorb the impact forces during exercise and played a cushioning role thanks to the unique structure formed by plantar dermis [20]. Zhang et al. constructed a finite-element 3D model of the ostrich foot using reverse engineering technology and experimentally simulated the plantar pressure distribution of the ostrich foot. Their findings indicated that the pressure on the middle part of toe III of the ostrich foot was the minimum [21]. Wang Haitao from Jilin University extracted the structure of the ostrich foot pad and established a finite-element analysis model accordingly. The modal analysis results suggested that the foot pad was mainly deformed in the form of bending at both ends, which was beneficial to protect the ostrich foot musculoskeletal system from impact damage [22]. When studying the mechanism of metatarsophalangeal joint, they also found that the ostrich metatarsophalangeal joint not only achieved compression and cushioning, but also provided energy return for forward motion [23]. Studies have shown that the metatarsophalangeal joint absorbs 63.3% of the total energy stored in the entire ostrich leg and plays an important role in cushioning and shock absorption [24]. Prof. Rui Zhang et al. studied the mechanism of the ostrich's sand crossing and found that the metatarsophalangeal joint could promote the rebound of the toe joint during the stirrups off the sand phase when the ostrich was running fast [25]. By observing and analyzing the microstructure of tendons, Zhang et al. suggested that the large number of collagen fibrous structures of tendons worked with the bones of the metatarsophalangeal joint for force transfer and play a role in the energy storage and cushioning of the metatarsophalangeal joint [26].

The research objective of this paper is to develop a sports shoe with excellent cushioning performance by coupling material and structural design in the limited thickness space of the midsole (including forefoot and heel) of sports shoes using engineering bionic technology. Inspired by the excellent cushioning and shock absorption characteristics of the ostrich foot, a new type of bionic cushioning unit for the sole was developed based on the mechanism of the toe pad and the metatarsophalangeal joint, with a targeted functional bionic design for forefoot and heel areas. The cushioning performance of the bionic sole was evaluated and analyzed by testing the cushioning performance of the sole. Therefore, this paper provides innovative reference and research ideas for the design and development of sports shoes with good cushioning performance.

2. Materials and Methods

2.1. Analysis of Key Bionic Parts of Ostrich Foot

2.1.1. The Skeletal–Tendon of the Ostrich Foot Metatarsophalangeal Joint

The metatarsophalangeal joint of the ostrich foot remains off the ground throughout locomotion and exhibits an up-and-down floating compression and rebound posture during foot contact, similar to a spring [27]. The model of the skeletal–tendon at the metatarsophalangeal joint of the ostrich foot was reconstructed in reverse by medical imaging scanning, and the schematic diagram of the transmission of the metatarsophalangeal joint forces was analyzed according to the inverse reconstruction model and the anatomical model. It can be seen from the structural features inside the ostrich foot in Figure 1 that the tendon was below the skeletal and extended upward from the tip of the foot in parallel with the skeletal, forming a structural combination of tendon, tendon sheath, and bone joint features at the metatarsophalangeal joint. The presence of tendon restricted the flipping of the metatarsophalangeal joint and kept it in an elevated position. The metatarsophalangeal joint area rotated and was pressed downward during movement, transferring force to the tendon below the joint through tendon sheath. In order to maintain joint elevation, downward pressure was converted into tension force within the tendon. According to studies, at the position of the metatarsophalangeal joint, the stiffness at both ends of the tendon is less than that in the middle, giving the tendon elastic deformation capacity that facilitates the absorption of impact energy by stretching the tendon when the joint is subjected to downward force. Therefore, the tendon plays a cushioning and shock-absorbing role in the metatarsophalangeal joint [28].

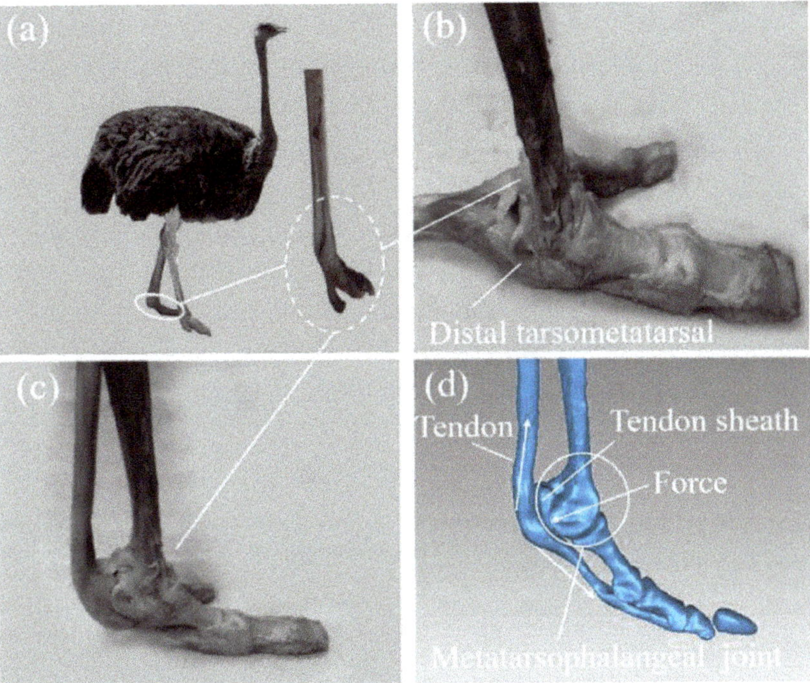

Figure 1. Schematic diagram of the transmission of force in the metatarsophalangeal joint. (**a**) Ostrich foot; (**b**) Metatarsophalangeal joint after fat clearance and tendon separation; (**c**) Metatarsophalangeal joint after fat clearance and tendon recovery; (**d**) 3D model of ostrich foot.

We observed tissue sections of the tendons in the metatarsophalangeal joint area to further verify their elastic deformability, as illustrated in Figure 2. In addition, the properties of the material from the tissue structure of the tendon biomaterial were analyzed. The results showed that collagen fibers at the proximal and distal locations of the metatarsophalangeal joint (the less stiff part of the material) were arranged in a wavy pattern in parallel in the longitudinal direction of the tendon. The wavy collagen fibers endowed the tendon with longitudinal tensile deformation and enabled it to convert the kinetic energy of the joint into the elastic potential energy of the tendon by elastic deformation under joint pressure [29].

According to the above analysis, the ability of the metatarsophalangeal joint of the ostrich foot to perform its cushioning function depended on its structure and form of material assembly. During movement, the metatarsophalangeal joint transferred force through the structural assembly of hard bone and soft tendon on the one hand and absorbed impact kinetic energy by taking advantage of the tendon's elastic deformation on the other.

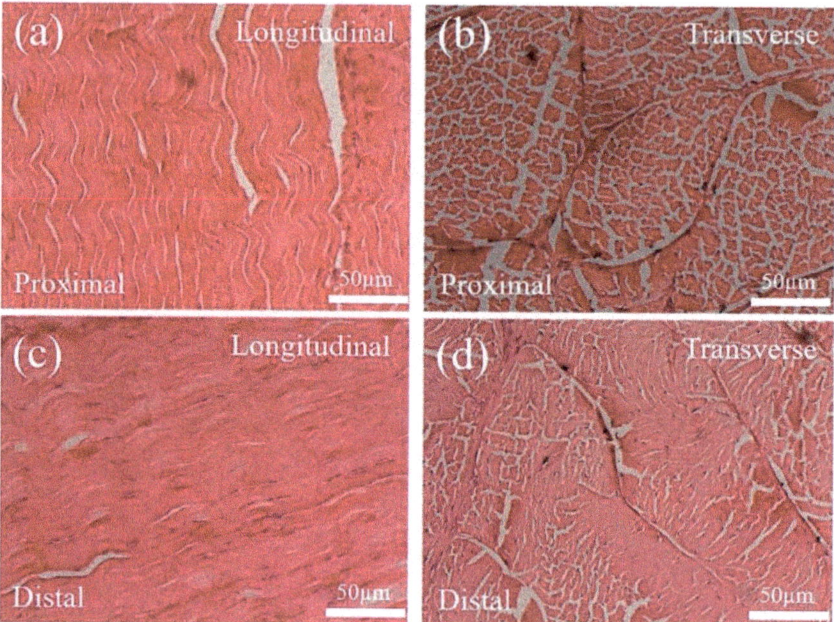

Figure 2. Longitudinal (**a**) and transverse (**b**) views of the third toe tendon at the superior end of the metatarsophalangeal joint.and Longitudinal (**c**) and transverse (**d**) views of the third toe tendon at the lower end of the metatarsophalangeal joint.

2.1.2. Pad–Fascia of Ostrich Toe

The third toe of the foot with a relatively wide foot pad was selected as the anatomical sample, and the tissue structure of the foot pad was separated using gross anatomical methods. The sample shown in Figure 3 was extracted from the right foot of male ostriches aged 2 years in the ostrich captive colony site in Team 7 at Qingfeng Farm, Hulin City, Heilongjiang Province, China. In order to avoid damage to the overall structure of the foot pad, a cutting process was performed at the dorsal skin of the toe by cutting the skin layer along the position marked in white with medical scissors, and the skin from the bone at the skin break was slowly peeled to remove the bone and toenail using a medical scalpel. By observing the dissected toe pads and testing the material hardness, we found that the foot pad consisted of three sections, namely, the outer layer of skin, the inner layer of toe pads, and the middle layer of fascia. The footbed shows the structure and material assembly characteristics of the outer layer covering the inner layer and the material hardness from the outer layer to the inner layer from hard to soft.

In addition, the microstructural information of the toe pad and the interface between the toe pad and the fascia was obtained by tissue section observation, and the microstructures of the two biomaterials were compared to analyze the functional roles of the two materials. As can be seen in Figure 4, the fat in the stained state of the toe pad is marked in white, and a large amount of fat inside was separated into numerous individual small compartments by pink collagen fibers. Fat can play a good role in energy absorption and protect the body from mechanical damage. There was a large amount of fat in the toe pad, forming numerous small, energy-absorbing cushioning units. The microstructure of the toe pad was similar to that of the microporous cushioning foam used in the midsole of sports shoes. Numerous hollow micropores emerged inside microporous foam material in the foaming process, and then gas in the holes was compressed and absorbed impact energy by taking advantage of the material's deformation caused by force.

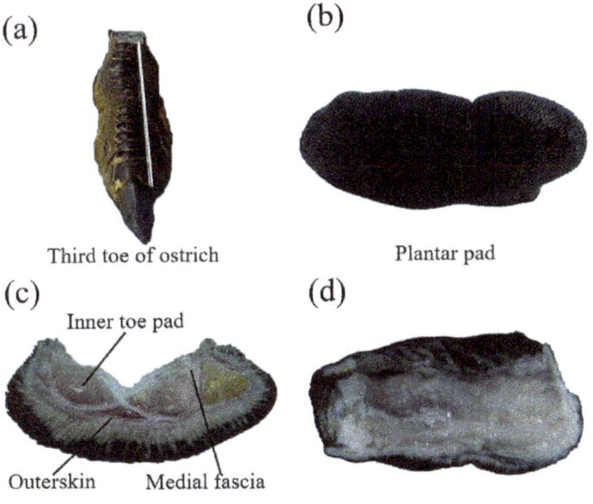

Figure 3. Anatomical diagram of the toes. (**a**) Anatomy of ostrich toes; (**b**)Topography of toe pad bottom surface; (**c**) Section of the third toe pad; (**d**) Overall overhead view.

The fascia is embodied as a septum between the skin and the toe pad, enveloping the outer layer of the toe pad, and its microscopic organization is significantly different from that of the toe pad, which contains numerous fatty compartments, as shown in Figure 5. As can be seen from the figure, stained elastic fibers exhibited a translucent wavy structure, and the fascia consisted of a large number of elastic fibers, among which a small number of collagen fibers were distributed. According to the difference in the tissue structure of the interface biomaterials between fascia and toe pad, the two biomaterials were speculated to have different functions. The wavy elastic fibers were arranged in parallel into fascia so that the fascia, a biological material, have a certain elastic expansion capacity. The elastic wrapping effect of fascia on the toe pad could ensure that the toe pad could maintain a stable shape during deformation and energy absorption.

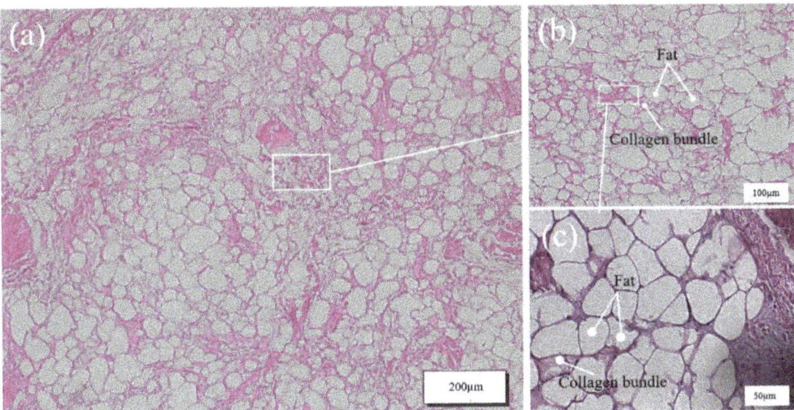

Figure 4. Diagram of a toe pad tissue section. (**a**) Longitudinal section of toe pad; (**b**) Enlarged view of the white box in figure (**a**); (**c**) Enlarged view of the white box in figure (**b**).

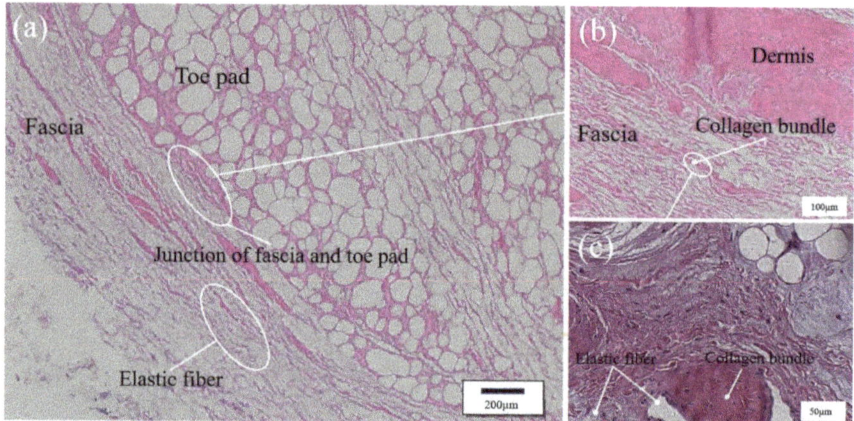

Figure 5. Diagram of a fascia tissue section. (**a**) Longitudinal section of fascia; (**b**) Enlarged view of the oval white box in figure (**a**); (**c**) Enlarged view of the oval white box in figure (**b**).

2.2. Bionic Design

Based on the skeletal–tendon structure and material assembly characteristics of the ostrich metatarsophalangeal joint, the forefoot cushioning unit model was designed by replacing the bone with a rigid material to transfer the force to a soft elastic material. The forefoot cushioning unit, as shown in Figure 6, was composed of an outer frame and sandwiched elastic energy-absorbing elements. The inner part of the outer frame was arranged in parallel with a semi-circular convex structure, and the upper frame and the lower frame with a convex structure were staggered. Each individual semi-cylindrical projection interacted with interlayer elastic energy-absorbing element to form a skeletal joint–tendon bionic shock-absorbing structure. When the unit was impacted, the semi-cylindrical projections of the outer frame transferred force to the sandwich elastic energy-absorbing element when the unit was subjected to impact and then converted the kinetic energy of the impact into its own elastic potential energy by means of the elastic deformation of the elastic energy-absorbing element itself.

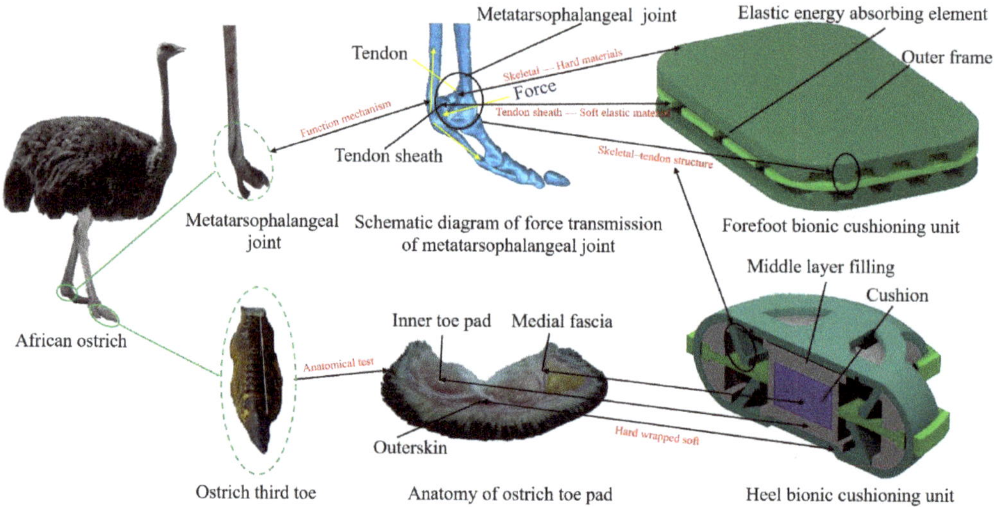

Figure 6. Schematic diagram of forefoot and heel cushioning unit model structure.

In order to improve the cushioning performance of the heel, a bionic cushion assembled by the toe pad and fascia inside the foot pad was added on the basis of forefoot cushioning unit. The heel bionic cushioning unit model consisted of an outer frame, an elastic energy-absorbing element, a cushion, and a middle layer filling, whose structural features are shown in Figure 6. The figure shows that the outer frame of hard material was distributed with an extended rocker structure with an included angle of 45° with the frame plate surface in four directions. The extended rocker was staggered up and down, transferring force to the soft, elastic energy-absorbing element placed between the upper and lower rocker as a bionic cushioning structure. The cushion pad was made from the inner layer material (purple part in Figure 6) selected from the traditional microporous foam material of the midsole to replace the fat compartment material of the toe pad, and the outer layer elastically wrapped around the inner layer and integrated with the unit's interior as a one-piece structure. The same silicon rubber material (white part in Figure 6) was selected for processing. The cushion through the inner layer of deformation energy absorption and the outer layer of elastic wrapping to achieve the toe pad–fascia cushioning function bionic and skeletal–tendon design ideas combined, applied to the heel bionic cushioning unit prototype design.

2.3. Sample Preparation

2.3.1. Principles of Sample Size Design

Taking the traditional foam sole as the base, the cushioning unit was embedded inside the midsole to design and produce physical bionic cushioning unit. The contour size of the bionic cushioning unit was designed with reference to the pre-selected basal to ensure that the unit was fully embedded inside the sole. As shown in Figure 7, the cushioning unit was arranged in the rectangular and circular areas at the forefoot and heel, respectively, whose contour size did not exceed that of the outer edge of the basal. Due to the inconsistent thickness between different areas of the midsole, the thickness of the forefoot and heel cushioning units should not exceed 25 mm and 15 mm, respectively.

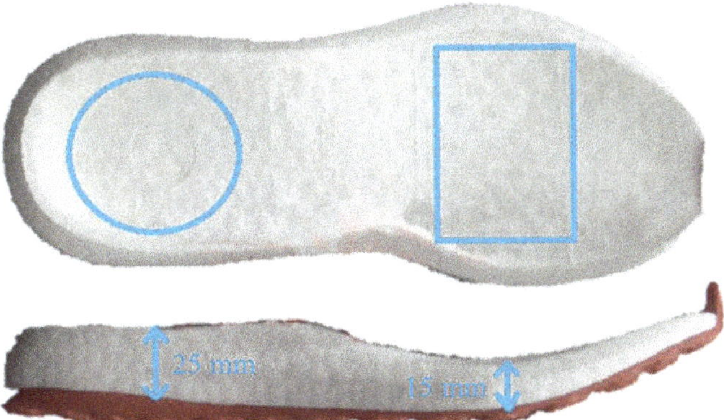

Figure 7. The profile of shoe base.

2.3.2. The Machining Process of the Bionic Cushioning Unit

In order to better process the physical parts of the cushioning unit and reflect the excellence of structural and material coupling bionic designs, the cushioning unit parts were assembled and processed with traditional conventional materials, and then the bionic cushioning shoe sole unit was physically produced. The outer frame of the unit should have certain stiffness to ensure the stability of the unit's structural support. Furthermore, in order to prevent the cushioning unit in the midsole from being too stiff to make the foot uncomfortable, the frame material also needed to have certain elastic deformation capacity. Therefore, relatively soft PU rubber material was selected for producing the raw

TPU printing wire material with a stiffness range of 80–95 HA TPU, 3D printing outer frame, and elastic energy-absorbing components.

The outer frame of the forefoot bionic cushioning unit was printed using a 3D printer with silk fusion technology, with a maximum size of $150 \times 150 \times 150$ mm, a nozzle diameter of 0.4 mm, and printing accuracy of 0.1 mm. TPU printing consumables with a diameter of 1.75 mm were selected, and the printhead temperature was set to 230 °C and the hot bed temperature to 30 °C. To avoid plugging in the molding process, the thickness of printing layer was set to 0.2 mm, and printing speed to 40 mm/s. The machining and assembly process of the bionic forefoot cushioning unit is shown in Figure 8a. The entire unit was 12 mm thick, 54 mm wide, and 85 mm long, whose contour dimensions met the dimensional requirements for embedding in the forefoot area, as shown in Figure 8a. The heel bionic cushioning unit frame was processed in the same way. To facilitate the addition of the internal structure of the heel bionic cushioning unit, the outer frame model of heel bionic cushioning unit was split main body and upper cover, which were printed separately. The whole unit was filled with the cushion pad and the outer cushion pad, and the silicone sheet formed with the stiffness of 25–30 HA was purchased for manual cutting. The machining and assembly process of the bionic heel cushioning unit is shown in Figure 8a. The finished heel cushioning unit with a profile diameter of 56 mm and an overall unit thickness of 22 mm is shown in Figure 8b.

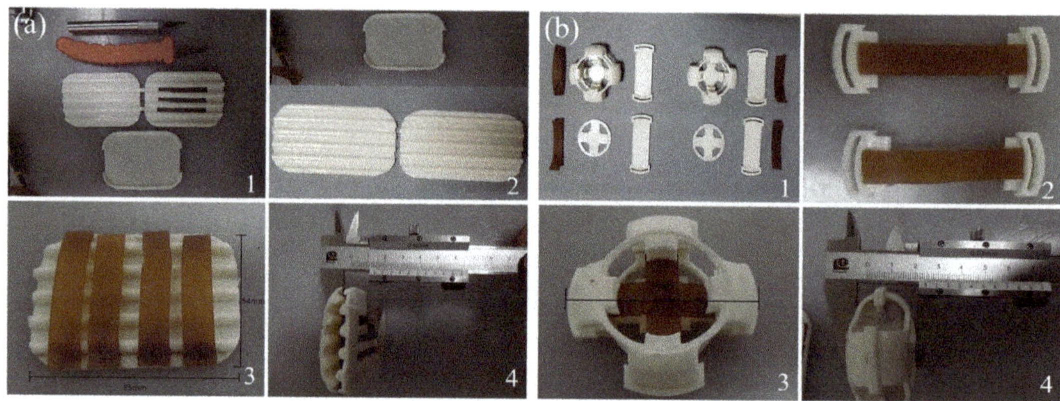

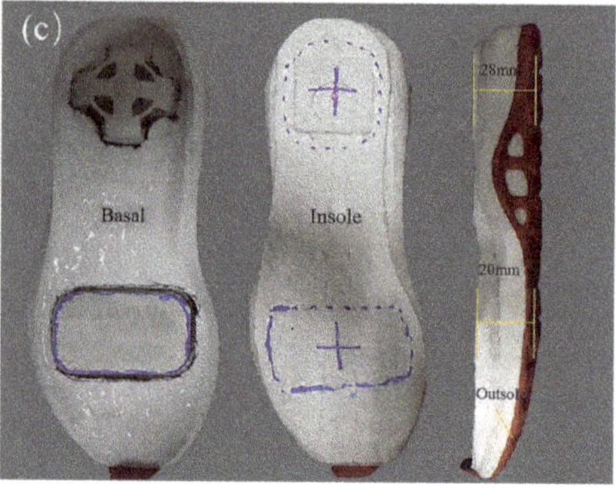

Figure 8. The machining process and complete assembly diagram of cushioning sole. ((**a**) Forefoot cushioning unit. (**b**) Heel cushioning unit. (**c**) The assembly diagram of cushioning sole.)

The bionic cushioning sole was composed of an outsole, midsole, and insole. With the outsole and midsole as the base, the forefoot and heel units were grooved according to the contour size of the cushioning unit. Then, the cushioning unit was inserted inside the groove hole of the midsole, and the glue particularly used for repairing athletic shoe soles was evenly applied to the upper surface of the midsole and the cushioning unit, and finally to the insole to generate bionic cushioning sole. The thickness of forefoot and heel buffer zones was, respectively, measured to be 20 mm and 28 mm, as shown in Figure 8c. The conventional control sole was produced using a homogeneous material without any structure, which was the same as cushioning unit, and its thickness at the forefoot and heel was the same as that of the bionic cushioning sole.

2.4. Impact Test

2.4.1. Test Devices

The test device consisted of impact test bench, data collector, and PC data display port, as shown in Figure 9. The impact test bench was composed of impact hammer, light bar, and bearing table. According to the requirements of GB/T 30907, national standard for testing shock absorption in sports shoes, the diameter of the disc where the impact head was in contact with the specimen was 45 mm, and the mass of the impact hammer was 8.5 kg after the addition of the counterweight. The impact head bolted onto the beam was moved vertically up and down along the parallel polished rods during tests.

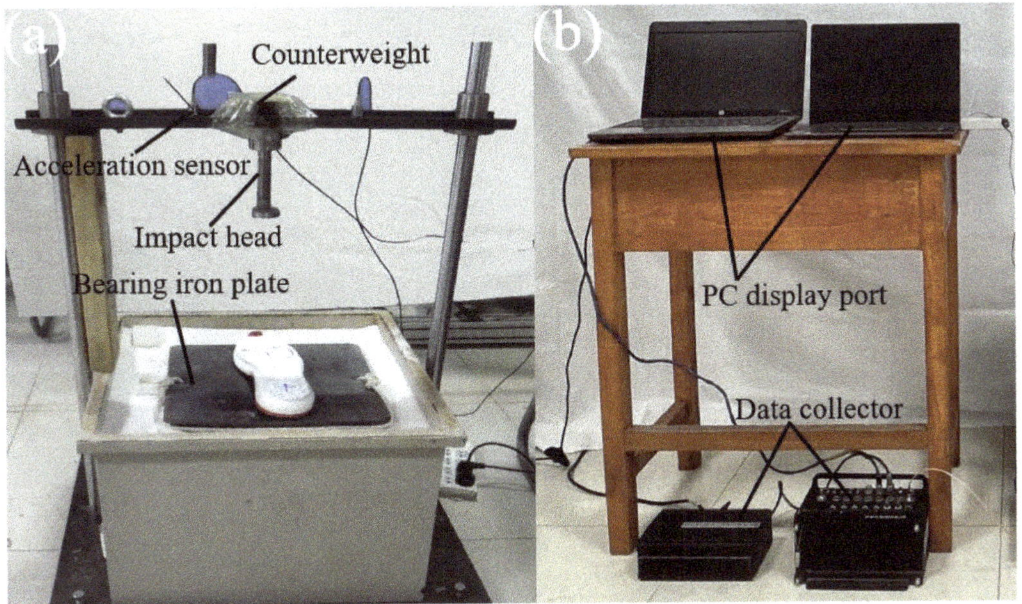

Figure 9. Test device. (**a**) Impact test bench; (**b**) Data acquisition system.

2.4.2. Test Media

The solid ground was simulated using 45 steel plates in the cushioning performance tests.

2.4.3. Test Process

A total of six high marks were made within the range of 4–14 cm. The impact test acquisition system was connected with the acceleration sensor, and the fastening of the acceleration sensor was checked to avoid the instability of data acquisition caused by loose connection. The connecting part of the polishing rod of the impact hammer and the linear bearing were lubricated with engineering lubricating oil. A height scale of 4 cm was pre-selected to adjust the vertical distance between the impact hammer and the test sole.

The impact test was carried out on the forefoot and heel areas of the bionic cushioned sole and conventional control sole, respectively, at the impact heights of 4 cm, 6 cm, 8 cm, 10 cm, 12 cm, and 14 cm with an impact hammer of 8.5 kg and the impact energies of 3.3 J, 5 J, 6.6 J, 8.33 J, 10 J, and 11.6 J, respectively. The test was repeated at least five times under each impact energy to ensure that all five groups of valid data were recorded.

2.4.4. Data Acquisition and Processing

A data-acquisition and -control system consisting of acceleration sensors (EA-YD-186, 10.07 PC/ms^{-2}), a mobile data acquisition device (MDR-81), and a personal computer was employed. The signal was acquired using two acceleration sensors during tests to ensure data accuracy. The acceleration sensors were fastened to the left and right ends of the impact hammer by M5 bolts. The data-acquisition device had a sampling frequency of 256 Hz and an acquisition time of 3 s.

In the software display interface, the data curves were observed to eliminate the data with large variability among the five sets of valid data at each impact energy. Three sets of stable data were exported and saved, each of which contained acceleration data recorded by two sensors. A total of six data samples were retained. The peak negative acceleration values in the data samples were extracted and averaged after the exclusion of the extreme value. Then, the curves were plotted accordingly.

3. Results

3.1. A Comparative Study of Shoe Sole Cushioning Performance

The peak negative acceleration curves of the bionic cushioning sole and conventional control sole in the forefoot and heel areas under impact energies of 3.3 J, 5 J, 6.6 J, 8.33 J, 10 J, and 11.6 J are shown in Figure 10. The peak negative acceleration was the maximum acceleration when the impact hammer collided with the sole. The peak negative acceleration of the impact hammer was used to measure the cushioning performance of the sole. When the soles with the same thickness were under the same impact energy, the smaller the negative acceleration peak value obtained on the impact hammer, the better the cushioning performance of the sole. The peak negative acceleration is represented by the g value in the plot and calculated as the ratio of the acceleration value of impact hammer to the gravitational acceleration.

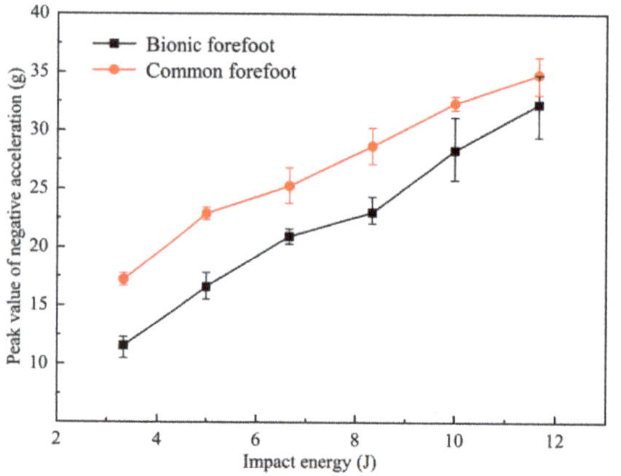

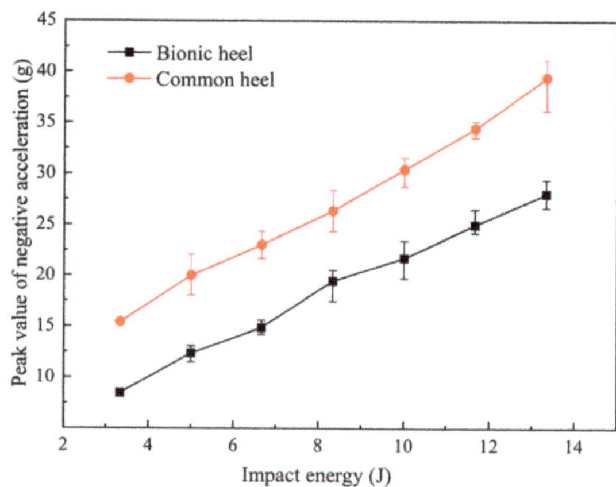

Figure 10. The average negative acceleration peak curve of a bionic or ordinary sole.

As can be seen from Figure 10, with an increase in impact energy, the peak negative acceleration in the forefoot and heel areas of the bionic cushioned sole approximately increased linearly but lower than that of the control sole in case of the same impact energy. According to the above analysis, the cushioning performance of bionic soles was effectively improved after the addition of forefoot and heel bionic cushioning units compared with the conventional control soles made of ordinary homogeneous materials.

3.2. Analysis of Additional Absorption Impact Increment and Growth Rate

Chiu proposed [30] that the relationship between the peak negative acceleration and the impact energy could be derived by fitting a linear equation. The linear regression equations and coefficients of determination of the equations obtained by fitting the four data curves in Figure 10 are shown in Table 1, where a represents the average value of the peak negative acceleration obtained from the impact hammer and E represents the impact energy acting on the sole of the shoe when the impact hammer falls at a specific height.

Table 1. Linear regression equation of mean negative acceleration peak (a) and impact energy (E).

Shoe Sole Area	Shoe Sole Type	
	Bionic	**Ordinary**
Forefoot	$a_1 = 2.41E + 4.03 \ (R^2 = 0.99)$	$a_2 = 2.06E + 11.46 \ (R^2 = 0.99)$
Heel	$a_3 = 1.98E + 2.14 \ (R^2 = 0.99)$	$a_4 = 2.23E + 8.25 \ (R^2 = 0.99)$

The same impact energy was applied to the cushioned sole and conventional control, the difference Δa_{m-n} in the negative acceleration peak value represents the impact energy additionally absorbed by the bionic sole, that is, compared with the ordinary sole, the additional impact-absorption increment of the bionic sole was calculated by Formula (1). The additional absorption impact growth rate of the bionic sole relative to that of ordinary sole is expressed by F_{m-n}, which can be calculated by Formula (2).

$$\Delta a_{m-n} = a_m - a_n = (\alpha_m - \alpha_n)E + (\beta_m - \beta_n) \tag{1}$$

$$F_{m-n} = \frac{\Delta a_{m-n}}{a_m} \tag{2}$$

where a_m and a_n are the peak negative acceleration obtained by the collision between impact hammer of ordinary soles and bionic soles; α and β are the linear regression equation coefficient; E is the impact energy of impact hammer at a specific height acting on the sole of the shoe.

According to the linear regression equation of impact energy and negative acceleration peak value, the calculation equations of additional absorption impact increment and growth rate in the forefoot and heel areas of the sole were obtained, as shown in Table 2.

Table 2. Calculation equation for absorption impact (g) and absorption impact growth rate (%) in the sole area.

Shoe Sole Area	Variable	
	Additional Impact Absorption (J)	**Percentage of Additional Absorbed Impact (%)**
Forefoot	$\Delta a_{2-1} = -0.35E + 7.43$	$F_{2-1} = (-0.18E + 7.43)/(2.06E + 11.46)$
Heel	$\Delta a_{4-3} = 0.25E + 6.11$	$F_{4-3} = (0.25E + 6.11)/(2.23E + 8.25)$

The six impact energy values were input into formulas to plot the curve, as shown in Figure 11. By comparing the impact-absorption increment in the forefoot and heel areas, we found that additional impact-absorption increment in the heel area was higher than that in the forefoot area under all impact energies. With an increase in impact energy, the impact-absorption increment in the heel area slowly increased, while that in the forefoot area gradually decreased. In the impact energy range of 3.3–11.6 J, the additional absorption impact range in the forefoot and heel areas was 3.48–6.40 g and 6.94–9.02 g, respectively.

With an increase in impact energy, the ratio between the two decreased. In the impact energy range, additional impact absorption in the forefoot and heel areas increased by 9.83–34.95% and 26.34–44.29%, respectively.

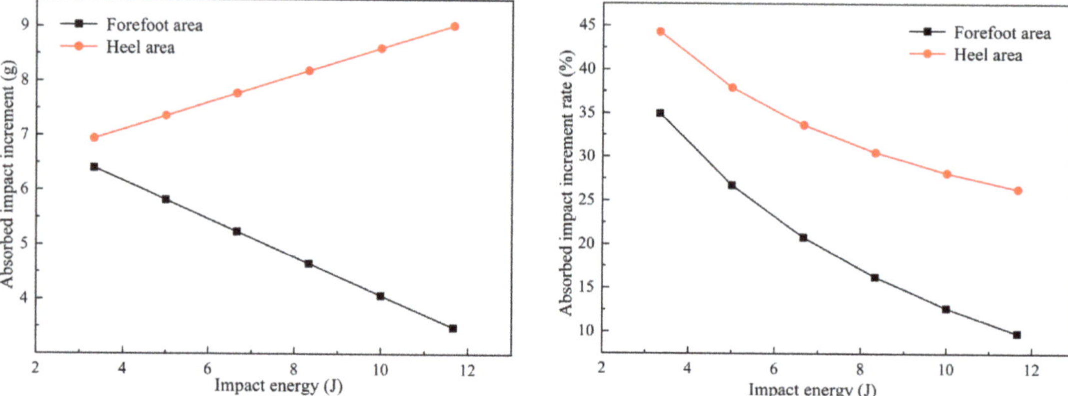

Figure 11. The curve of the bionic sole's additional absorption impact increment (g) and additional absorption impact growth rate (%).

3.3. Analysis of Additional Absorbed Impact Energy Increment and Growth Rate

It was found, from the energy-absorption effect in the forefoot and heel areas relative to the conventional control sole, that the negative acceleration peak of bionic sole was the same as that of ordinary sole, indicating that both have the same cushioning ability, as shown in Formula (3), where ΔE_{n-m}, the difference in the impact energy between the bionic sole and conventional control sole represents the additional energy absorbed by the bionic sole relative to the ordinary sole and can be calculated by Formula (4); H_{n-m}, the ratio of the impact energy increment to the initial energy of the impact hammer, is the proportion of the additional energy absorbed by the bionic sole, which can be calculated by Formula (5).

$$a_m = \alpha_m \times E_m + \beta_m = \alpha_n \times E_n + \beta_n \tag{3}$$

$$\Delta E_{n-m} = E_n - E_m = \frac{(\alpha_m - \alpha_n)E_n + (\beta_m - \beta_n)}{\alpha_m} \tag{4}$$

$$H_{n-m} = \frac{E_n - E_m}{E_n} = \frac{(\alpha_m - \alpha_n)E_n + (\beta_m - \beta_n)}{\alpha_m E_n} \tag{5}$$

where a_m and a_n are the peak negative acceleration obtained by the collision between impact hammer of ordinary soles and bionic soles; α and β are the linear regression equation coefficient; E is the impact energy of impact hammer at a specific height acting on the sole of the shoe.

The above formulas were integrated to calculate the additional absorbed energy and additional absorbed energy rate of the soles, forefeet, and heels, as shown in Table 3.

Table 3. Calculation equation of the additional absorbed impact energy (J) and additional absorbed energy growth rate (%) of the bionic sole.

Shoe Sole Area	Variable	
	Additional Absorbed Energy (J)	**Percentage of Additional Absorbed Energy (%)**
Forefoot	$\Delta E_{1-2} = (-0.35E_n + 7.43)/2.06$	$H_{1-2} = (-0.35E_n + 7.43)/2.06E_n$
Heel	$\Delta E_{3-4} = (0.25E_n + 6.11)/2.23$	$H_{3-4} = (0.25E_n + 6.11)/2.23E_n$

The absorbed energy increment of the sole was calculated by the formula in Table 4, and the absorbed energy increment curve of the sole is shown in Figure 12. It can be seen from the curve that as impact energy gradually increased from 3.33 J to 11.46 J, the additional absorbed energy increment in the forefoot area linearly decreased, and the absorbed energy growth rate gradually decreased with an increase in impact energy. The additional absorbed energy was 1.39–2.59 J, and the proportion of additional absorbed energy reached 10.65–43.84%. The energy-absorption increment and the proportion of absorbed energy in the heel area were higher than those in the forefoot area, and the additional absorbed energy increment linearly increased. The proportion of additional absorbed energy gradually decreased with an increase in impact energy. In the entire impact energy range, the additional absorbed energy in the heel area was 3.51–4.56 J, and the proportion of additional absorbed impact energy was 28.1–51.29%.

Table 4. Comparisons of Impact Testing Methods Used in Previous Studies and in the Present Study.

Investigator	Shoe Sole Area	The Range of Impact Energy (G)	Additional Impact Absorption (g)	Percentage of Additional Absorbed Impact (%)	Additional Absorbed Energy (J)	Percentage of Additional Absorbed Energy (%)	Peak Value of Negative Acceleration (g)
Yu HB [30]	Forefoot	2.35–5.88	0.016–0.019	7.15–12.49	0.396–0.466	7.92–16.88	1.05–2.40
	Heel	2.35–5.88	0.010–0.019	7.06–8.14	0.246–0.447	7.59–10.45	1.10–2.30
Zhou S [31]	Forefoot	2.35–5.88	1.5–5.9	0.221–0.243	0.31–1.22	0.142–0.210	6–16
	Heel	2.35–5.88	1.6–5.3	0.235–0.237	0.33–1.30	0.157–0.210	5–16
Hung HT [32]	Forefoot	-	-	-	-	-	-
	Heel	2–6	-	5.0–7.3	0.52–0.60	12–33	10–17
Xiao Y [33]	Forefoot	-	-	-	-	-	-
	Heel	6–10	-	-	-	-	23–32
Present study	Forefoot	3.3–11.6	3.48–6.40	9.83–34.95	1.39–2.59	10.65–43.84	12–31
	Heel	3.3–11.6	6.94–9.02	26.34–44.29	3.51–4.56	28.10–51.29	6–27.50

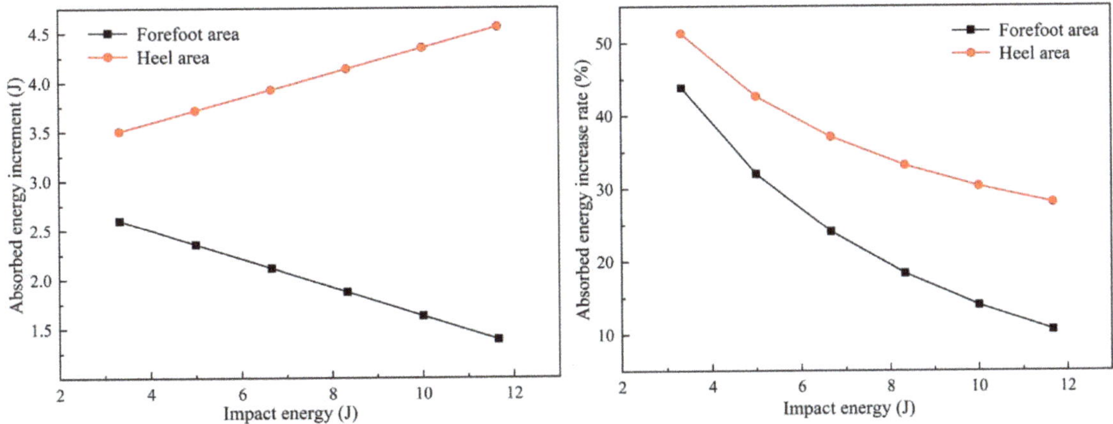

Figure 12. The curve of the bionic sole's additional absorption impact energy increment (g) and additional absorption impact energy growth rate (%).

4. Discussion

This study investigated the cushioning effect of bionic soles with bionic cushioning units versus conventional ordinary soles at different impact energies. The results of this study provide ideas for the development of cushioned sports shoes. Our research suggests that the cushioning performance of the bionic cushioned sole in the forefoot and heel areas is better than that of conventional ordinary shoes. At the same impact energy, the bionic cushioning sole absorbs 3.48–6.40 g more impact than the conventional control sole in the forefoot area, with a proportion of additional impact absorption of 9.83–34.95%. The additional absorption impact increment in the heel area is 6.94–9.02 g, and the additional absorption impact growth rate is 26.34–44.29%. Under the same conditions, the heel area outperforms the forefoot area in terms of energy absorption.

We also compared other soles and found that different athletic shoe soles were tested using a single impact energy of 2–10 J in previous studies (Table 4). As different shoes were tested in each study, only the results for the shoes with the best cushioning performance are shown in the table. By comparing the experimental data with those of Yu [30] and Zhou [31], we found that the bionic sole studied in this paper outperformed ordinary soles in terms of both impact absorption and impact energy absorption, which indicates its better cushioning performance. In addition, by comparing the cushioning data obtained at the minimum impact energy in this work with those obtained at the maximum impact energy in the studies of Yu and Zhou, the same conclusion as above can be obtained, so the above analysis results can be ruled out as being caused by the difference in impact energy. The peak negative acceleration is much larger than the other two because the impact energy of this work was much larger than that in the other studies. This study also found that the ratio of impact absorption and energy absorption in the forefoot and heel areas of the bionic sole gradually decreased as the impact energy increased, indicating that the cushioning performance of the bionic sole would be weakened with an increase in impact energy. This is consistent with the findings of the other two studies and proves the validity of this study. In addition, by comparing the data with the studies of Huang [32] and Xiao [33], we found that the maximum peak negative acceleration of this work was the smallest, which also indicated that the bionic sole of this study had better cushioning performance.

Some limitations of this study should be mentioned. Firstly, the bionic cushioning midsole combining the bionic cushioning unit and the midsole matrix of the shoe was not optimized. The coupled cushioning effect of the midsole and the bionic cushioning unit should be considered comprehensively to further investigate the improvement in the cushioning performance of the sports shoes. Secondly, cushioning performance is only one of the functional indicators to measure athletic shoes. On the basis of research on cushioning performance, the balance between multiple performances should be comprehensively considered so to ensure a better cushioning performance and wearing experience from sports shoes. Finally, we have only considered impact tests in this work, and human tests should be added to further increase its credibility.

5. Conclusions

Based on the excellent cushioning performance of the ostrich foot and inspired by the structure and material assembly features of the ostrich foot's metatarsophalangeal skeletal–tendon and the ostrich toe pad–fascia, a functional bionic cushioning unit for the midsole (including the forefoot and heel) area of athletic shoes was designed using engineering bionic technology. The bionic cushioning unit was processed based on the bionic design model, and then shoe soles were tested under impact energies ranging from 3.3 J to 11.6 J for drop hammer impact. Finally, the test results were compared with those on the conventional control sole with the same size. The results show that the cushioning performance of the sole can be effectively improved by adding a bionic cushioning unit inside the midsole. The cushioning performance of the bionic cushioning unit in the heel area has been improved significantly, while that of the bionic cushioning unit in the forefoot area still needs to be further optimized and improved.

Author Contributions: Conceptualization, R.Z. and L.Z.; data curation, H.Y.; formal analysis, G.Y. and H.Y.; methodology, G.Y. and H.Y.; resources, R.Z.; software, L.Z.; validation, J.L. and W.-H.T.; writing—original draft, L.Z. and Q.K.; writing—review and editing, R.Z. and H.Y. All authors have read and agreed to the published version of the manuscript.

Bioengineering **2023**, *10*, 1

Funding: We thank the support of the Science and Technology Development Planning Project of Jilin Province of China (No. 20220101014JC), the Science and Technology Development Planning Project of Fujian Province of China (No. 2022H6034), and the support of the Social Science Foundation Project of Fujian Province of China (No. FJ2022B023).

Institutional Review Board Statement: The study was conducted in accordance with the Declaration of Helsinki and approved by the Institutional Animal Care and Use Committee (IACUC, protocol number: 20140706) of Jilin University, China.

Informed Consent Statement: Informed consent was obtained from all subjects involved in the study.

Data Availability Statement: Data are contained within the article.

Conflicts of Interest: The authors declare no conflict of interest.

References

1. Lieberman, D.E.; Venkadesan, M.; Werbel, W.A.; Daoud, A.I.; D'Andrea, S.; Davis, I.S.; Mang'Eni, R.O.; Pitsiladis, Y. Foot strike patterns and collision forces in habitually barefoot versus shod runners. *Nature* **2010**, *463*, 531–535. [CrossRef] [PubMed]
2. Lin, S.; Song, Y.; Cen, X.; Bálint, K.; Fekete, G.; Sun, D. The Implications of Sports Biomechanics Studies on the Research and Development of Running Shoes: A Systematic Review. *Bioengineering* **2022**, *9*, 497. [CrossRef] [PubMed]
3. Li, J.; Gu, Y.; Lu, Y.; Wang, Y. Biomechanical study on the core technology of sports shoes. *China Sport Sci.* **2009**, *29*, 40–50.
4. Ruder, M.; Atimetin, P.; Futrell, E.; Davis, I. Effect of Highly Cushioned Shoes on Ground Reaction Forces during Running. *Med. Sci. Sports Exerc.* **2015**, *47*, 293–294. [CrossRef]
5. Chu, H.; Lin, P. Friction Analysis of the Sneakers Insole. *J. Biomech.* **2007**, *40*, S713. [CrossRef]
6. Ma, R.; Lam, W.-K.; Ding, R.; Yang, F.; Qu, F. Effects of Shoe Midfoot Bending Stiffness on Multi-Segment Foot Kinematics and Ground Reaction Force during Heel-Toe Running. *Bioengineering* **2022**, *9*, 520. [CrossRef]
7. Hoogkamer, W.; Kipp, S.; Frank, J.H.; Farina, E.M.; Luo, G.; Kram, R. A Comparison of the Energetic Cost of Running in Marathon Racing Shoes. *Sports Med.* **2017**, *48*, 1009–1019. [CrossRef]
8. Robbins, S.; Waked, E.; Gouw, G.J.; McClaran, J. Athletic footwear affects balance in men. *Br. J. Sports Med.* **1994**, *28*, 117–122. [CrossRef]
9. Clarke, T.; Frederick, E.; Cooper, L. Effects of Shoe Cushioning Upon Ground Reaction Forces in Running. *Int. J. Sports Med.* **1983**, *4*, 247–251. [CrossRef]
10. Hsu, C.-Y.; Wang, C.-S.; Lin, K.-W.; Chien, M.-J.; Wei, S.-H.; Chen, C.-S. Biomechanical Analysis of the FlatFoot with Different 3D-Printed Insoles on the Lower Extremities. *Bioengineering* **2022**, *9*, 563. [CrossRef]
11. Prajapati, M.J.; Kumar, A.; Lin, S.C.; Jeng, J.Y. Multi-material additive manufacturing with lightweight closed-cell foam-filled lattice structures for enhanced mechanical and functional properties. *Addit. Manuf.* **2022**, *54*, 102766. [CrossRef]
12. Liu, L. Study on the Effect of Sole Shock Absorption Structure on Foot Shock Absorption System. Ph.D. Thesis, Shaanxi University of Science and Technology, Xianyang, China, 2015.
13. Cui, L.N.; Zhang, X.; Shi, J. Research on functional shoe soles. In Proceedings of the 6th Seminar on Functional Textiles and Nanotechnology Applications, Beijing, China, 26 May 2006.
14. Cheng, P.; Qu, F. Overview of the progress of sports shoes technology. In Proceedings of the 12th National Sports Biomechanics Academic Exchange Conference, Taiyuan, China, 14 October 2008.
15. Delgado, T.L.; Kubera-Shelton, E.; Robb, R.R.; Hickman, R.; Wallmann, H.W.; Dufek, J.S. Effects of Foot Strike on Low Back Posture, Shock Attenuation, and Comfort in Running. *Med. Sci. Sports Exerc.* **2013**, *45*, 490–496. [CrossRef] [PubMed]
16. Hill, C.M.; DeBusk, H.; Knight, A.C.; Chander, H. Military-Type Workload and Footwear Alter Lower Extremity Muscle Activity during Unilateral Static Balance: Implications for Tactical Athletic Footwear Design. *Sports* **2020**, *8*, 58. [CrossRef] [PubMed]
17. Heiderscheit, B.C.; Chumanov, E.S.; Michalski, M.P.; Wille, C.M.; Ryan, M.B. Effects of Step Rate Manipulation on Joint Mechanics during Running. *Med. Sci. Sports Exerc.* **2011**, *43*, 296–302. [CrossRef]
18. Perl, D.P.; Daoud, A.I.; Lieberman, D.E. Effects of Footwear and Strike Type on Running Economy. *Med. Sci. Sports Exerc.* **2012**, *44*, 1335–1343. [CrossRef]
19. Alexander, R.M.; Maloiy, G.M.O.; Njau, R.; Jayes, A.S. Mechanics of running of the ostrich (*Struthio camelus*). *J. Zool.* **1979**, *187*, 169–178. [CrossRef]
20. El-Gendy, S.A.A.; Derbalah, A.; Abu El-Magd, M.E.R. Macro-microscopic study on the toepad of ostrich (*Struthio camelus*). *Veter-Res. Commun.* **2012**, *36*, 129–138. [CrossRef]
21. Zhang, R.; Wang, H.; Zeng, G.; Li, J. Finite element modeling and analysis in locomotion system of ostrich (*Struthio camelus*) foot. *Biomed. Res. India* **2016**, *27*, 1416–1422.
22. Zhang, R.; Han, D.; Yu, G.; Wang, H.; Liu, H.; Yu, H.; Li, J. Bionic research on spikes based on the tractive characteristics of ostrich foot toenail. *Simulation* **2020**, *96*, 713–723. [CrossRef]
23. Schaller, N.; D'Août, K.; Herkner, B.; Aerts, P. Phalangeal load and pressure distribution in walking and running ostriches (Struthio camelus). *Comp. Biochem. Physiol. Part A Mol. Integr. Physiol.* **2007**, *146*, S122. [CrossRef]

24. Rubenson, J.; Lloyd, D.G.; Heliams, D.B.; Besier, T.F.; Fournier, P.A. Adaptations for economical bipedal running: The effect of limb structure on three-dimensional joint mechanics. *J. R. Soc. Interface* **2010**, *8*, 740–755. [CrossRef] [PubMed]

25. Zhang, R.; Zhang, S.H.; Liu, F. Analysis of the mechanism of trans-sand movement in ostriches. In Proceedings of the Innovating Agricultural Engineering Science and Technology to Promote the Development of Modern Agriculture—2011 Annual Academic Conference of Chinese Society of Agricultural Engineering, Chongqing, China, 22 October 2011.

26. Zhang, R.; Han, D.; Luo, G.; Ling, L.; Li, G.; Ji, Q.; Li, J. Macroscopic and microscopic analyses in flexor tendons of the tarsometatarso-phalangeal joint of ostrich (*Struthio camelus*) foot with energy storage and shock absorption. *J. Morphol.* **2017**, *279*, 302–311. [CrossRef] [PubMed]

27. Zhang, S.; Clowers, K.; Kohstall, C.; Yu, Y.-J. Effects of Various Midsole Densities of Basketball Shoes on Impact Attenuation during Landing Activities. *J. Appl. Biomech.* **2005**, *21*, 3–17. [CrossRef] [PubMed]

28. Luo, G. *Bionic Research on Vibration Damping Adaptive Walking Foot Based on Structural Characteristics of Ostrich Metatar-Sophalangeal Joint*; Jilin University: Changchun, China, 2016.

29. Lieberman, D.E.; Warrener, A.G.; Wang, J.; Castillo, E.R. Effects of stride frequency and foot position at landing on braking force, hip torque, impact peak force and the metabolic cost of running in humans. *J. Exp. Biol.* **2015**, *218*, 3406–3414. [CrossRef] [PubMed]

30. Yu, H.-B.; Zhang, R.; Yu, G.-L.; Wang, H.-T.; Wang, D.-C.; Tai, W.-H.; Huang, J.-L. A New Inspiration in Bionic Shock Absorption Midsole Design and Engineering. *Appl. Sci.* **2021**, *11*, 9679. [CrossRef]

31. Zhou, S. Bionic Design and Analysis of Football Soles. *Comput. Digit. Eng.* **2020**, *48*, 1286–1289+1343.

32. Chiu, H.-T.; Shiang, T.-Y. Effects of Insoles and Additional Shock Absorption Foam on the Cushioning Properties of Sport Shoes. *J. Appl. Biomech.* **2007**, *23*, 119–127. [CrossRef]

33. Xiao, Y.; Hu, D.; Zhang, Z.; Pei, B.; Wu, X.; Lin, P. A 3D-Printed Sole Design Bioinspired by Cat Paw Pad and Triply Periodic Minimal Surface for Improving Paratrooper Landing Protection. *Polymers* **2022**, *14*, 3270. [CrossRef]

MDPI
St. Alban-Anlage 66
4052 Basel
Switzerland
Tel. +41 61 683 77 34
Fax +41 61 302 89 18
www.mdpi.com

Bioengineering Editorial Office
E-mail: bioengineering@mdpi.com
www.mdpi.com/journal/bioengineering